MINES

OF THE DILLSBURG, PENNSYLVANIA AREA,

And Other Related Enterprises

by

Robert F. Gayman

Organized in 1984, the Northern York County Historical and Preservation Society, Inc., (NYCHAPS) was formally incorporated in 1986, and qualified in 1992 as a §501(c)(3), non-profit, tax exempt organization. The Society is dedicated to the preservation of local history through cultural programs, publications, and historical acquisitions. NYCHAPS is proud to serve as one of Pennsylvania's cultural resources.

Published by

NORTHERN YORK COUNTY
HISTORICAL AND PRESERVATION SOCIETY, INC.

Post Office Box 355
Dillsburg, Pennsylvania 17019

First Edition October 1995
Second Edition May 1998

Library of Congress Catalog Card Number: 95-71413

ISBN 0-9648050-1-4

Printed in the USA by Copy Right Printers, Inc., Mechanicsburg, PA

PREFACE

The Borough of Dillsburg is in the northwestern corner of York County, Pennsylvania; the South Mountain (a part of the vast Appalachian Mountain chain) ends just west of the borough. From its earliest settlement, Dillsburg was primarily a farming community because the gentle rolling hills were quite fertile. Even today this area retains its farming heritage. However, during the last half of the nineteenth century and extending over a decade into the twentieth century, mining competed economically with farming.

The discovery of iron ore, white and black clay, copper and limestone deposits in the area brought about this change, Why and how did the deposits of these minerals get here? How rich were they? What became of the mining industry? The answer to some of these questions will unfold with the information compiled in the stories to follow.

In 1983, the 150th anniversary of the founding of Dillsburg was celebrated with many special events. As a result of the interest generated by a compiled history during the celebration, it was decided to organize and continue the work of historical preservation. To accomplish this, the Northern York County Historical and Preservation Society (NYCHAPS) was organized in June 1984.

At the NYCHAPS meeting on July 10, 1985, a tour of the mines in the Dogwood Hollow (west of the borough and at the base of the South Mountain) was conducted by Lee Karns. On October 3, 1985 a program on the geology of Northern York County was presented by Jay Lininger.

These two programs rekindled my interest in the mining industry. As boys, my next door neighbor, George Frederick, and I would fish in one of the mine holes along Ore Bank Road. It was probably the Big Pit of the Mumper-McCormick operation. We would catch catfish in this pond. Later it became Dillsburg's dump. George and I also hiked around the "Pinkerton" mine site on South Mountain which I found to be a pretty place. George, incidently, was the son of Paul and the nephew of Lewis Frederick, both of whom had worked in the Dogwood Hollow mines.

An interview with Lewis Frederick was taped by Lee Karns. This provided valuable information of recent history, which was unavailable in geological survey records.

Over four years ago when I volunteered at a NYCHAPS meeting to chair a committee to gather the mining information, I had no idea there were so many mining operations or how much had been written about them. The committee took many journeys into the past, visiting practically all the locations mentioned in this book, trying to imagine how things had been.

MORE INFORMATION WANTED

When this publication was prepared, the Committee researched all available sources. We believe there is more information, stories and photographs in private collections which we did not uncover. NYCHAPS would welcome additional information for its files.

MINE SITES ARE PRIVATE PROPERTY

After reading this publication on the mines of the Dillsburg area, some persons may be interested in visiting the mine sites. Please remember that all of the mine locations included in this publication are on private property. You must obtain permission from the current owner(s) before visiting these sites.

Robert F. Gayman
Dillsburg, Pennsylvania
1995

COVER PHOTOGRAPH

The Ore Washers provided the background for the company photo of the Pinkerton iron ore workers circa 1907-1908. The young lad in the foreground was Lewis Frederick who worked as a water boy. Foundation remains are visible on the north side of Tuckahoe Road.

Photo courtesy of Mrs. Lewis Frederick

ACKNOWLEDGEMENTS

The publication of this book is the culmination of a five year effort by the Northern York County Historical and Preservation Society, Inc. (NYCHAPS); and no effort that leads to a publication of this sort goes without the participation of many members and friends of the Society.

First and foremost, our thanks to Lee Karns for the notes, pictures and especially the taped interviews of former miners who shared their memories and experiences with him. This information served as the springboard for this publication.

Lena Kinter found an early source of pre-Civil War mining in Franklin Township. Lee Strayer provided many outstanding photos and stories about the Kuntz Limestone Quarry. Don and Gladys Baish obtained rare photographs from Mrs. Lewis Frederick, wife of a miner. Jill and John Hoffman granted permission to study the official minutes of the Dillsburg Vitrified Brick and Tile Company, which are in their collection. With NYCHAPS itself having purchased the correspondence of the brick company, these two resources gave us a valuable and interesting inside look at the machinations of establishing this enterprise, which due to many factors, especially the Panic of 1907, ultimately foundered.

Our appreciation extends to Bob Albert of Dillsburg for directing us to the Bender mine in Carroll Township; and to Norm Shelley, Jr., who showed us the site of the Altland mine on his parents' property. Perry Narten accompanied us to the location of three Monaghan Township mines. Our thanks to Jeri Jones, of Spring Grove, Pa., for providing information on the Dogwood Hollow mines and listing some of the unknown mines. Randy Watts of Carlisle, Pa., shared his knowledge about the railroads which served these mines. The restoration of the old photographs reproduced here is the result of the expertise of Ellis Wolfe of Dillsburg. Our thanks also extend to Mary Logan and Jay Young. Jay Lininger, well known and highly respected mineralogist, made an invaluable contribution by authoring the section on the Geological Formation. The publication effort also resulted in the acquisition by NYCHAPS of the "dinky rails" which were the gift of Robert Smith.

Many NYCHAPS members were very supportive in gathering information, pictures and leads to stories that could not have been told without these materials. Thanks to any and all those who helped in this project.

A very special thank you is extended to Bette Gayman, Bob's wife, who transcribed Bob's hand written drafts, while fulfilling all her many other responsibilities with distinction (including being a loving grandmother).

It is the Mines Committee members, listed here, through their enthusiasm and perseverance, who are responsible for the preparation of the report to the Society.

Mine Committee Members:

Robert F. Gayman, Chairman
M. James Hockenberry
George L. Jackson
Robert W. Watkins
Jacob L. Heisey
John O. Hoffman
Jay L. Lininger
Jack Winieski

While there were many participants in this project, by the agreement of all involved, Robert Gayman was the member without whose efforts this book would not be published. It was he who initiated the committee, guided it through all of its assignments, and who substantially prepared the report in its manuscript form. In a real sense, this is his book; and it is for this reason, and rightly so, that Robert Gayman is listed as the principle author of this publication.

The Committee's job ended when it presented its report to the Society. A Publication Committee was then formed to prepare the manuscript for publication. The Publication Committee consisted of the members of the Mine Committee along with Society President, Victor Neubaum; Treasurer, Chad Wagner; and member, Larry Klase. During this phase of the project, the Publication Committee, led by Jake Heisey, collectively put the manuscript in a form suitable for delivery to the printer.

When this book was published, the committee had exhausted all of the available sources known to it. It will be no surprise to NYCHAPS if more information, stories and photographs in private collections are revealed as the result of the interest in this book. We welcome any additional information that we can have on this topic of local historical interest.

Victor Neubaum,
President,
NYCHAPS
July, 1995

TABLE OF CONTENTS

SIDE BARS

SECTION I

MINES OF THE DILLSBURG, PENNSYLVANIA AREA,

and other related enterprises

by

Robert F. Gayman

CHAPTER ONE

IRON ORE MINES

The year was 1847. James Polk was president of the United States. The American Medical Association was formed and Joseph Pulitzer was born. The United States was at war with Mexico. The Post Office issued its first two stamps, a five-cent Ben Franklin and a ten-cent George Washington. And on a farm in southcentral Pennsylvania, iron ore was discovered near the Borough of Dillsburg in Carroll Township, York County, by Abraham Mumper. Soon after that, deposits were also found on the adjoining property owned by Abraham Mumper's brother, John.

After that second discovery, additional iron mining operations sprang up a few miles east of Dillsburg. Since the iron ore was located on farm land and the owners were unfamiliar with mining operations, leasing land to individuals or mining companies was common. After these leases were drafted, signed, sealed, and delivered, work would begin. The land owners were paid a royalty on each ton of iron ore removed. If the ore yield was less than anticipated, the common practice was to "throw up" the lease, with mineral rights returning to the land owner. It was not unusual to have two, three or more individuals or groups connected with one mine location.

Most of the iron ore found east of Dillsburg was magnetic. These were the first known deposits of magnetic ore west of the famed Cornwall Mines in Lebanon County, PA. Magnetic ore was also found in northern Monaghan Township, York County, adjacent to the Yellow Breeches Creek--less than a half mile from Grantham, Cumberland County. Magnetic ore was also found west of Wellsville in Washington Township, York County. Mining began there in 1872.

Non-magnetic iron ore (hematite or brown ore) was discovered in northern Franklin Township, York County, in the area known as Dogwood Hollow. Other similar deposits were found between the southeast side of South Mountain and Mountain Road.

ABRAHAM AND JOHN MUMPER

Abraham Mumper is credited with being the first to discover and mine magnetic iron ore on his farm in Carroll Township. The ore was found just two feet below the earth's surface. Abraham's brother John lived on an adjourning farm to the north and soon he became involved in the mining business.

Abraham was born January 16, 1786 at his father's home located at 1221 Mountain Road, Franklin Township (owned by John and Debbie Flanagan in 1994) several miles southwest of Dillsburg. His grandfather Michael Mumper, Senior, emigrated here from Germany and eventually owned over 1200 acres in Franklin and Carroll Townships. He is buried in Franklin Church Cemetery.

When Abraham Mumper was twenty-three he married Mary Lerew. In 1813 he purchased 200 acres from his parents and lived in a log house at the eastern edge of Dillsburg on Old York Road.

John Mumper was born on July 24, 1781. His first wife may have died in child birth. He married Jane Beelman from Monaghan Township in 1802. They moved to a log house east of Dillsburg located on Mumper Lane in 1803. John's father Michael Mumper, Junior, deeded 277 acres to him in 1811 and 1813. When Dillsburg was incorporated as a borough in 1833 in western edge of John's land formed the eastern boundary of the borough. It is believed that Ore Bank Road was created to serve the mining operations of the brothers and it was likely along the eastern edge of the land. Gibson's HISTORY OF YORK COUNTY reported the two brothers reaped a handsome profit from the mining operations.

Eventually both Mumper brothers had brick homes, Abraham at 341 South Baltimore Street (owned by Greg and Victoria Wonders in 1994) and John at 235 Mumper Lane (owned by Ted and Susan Shumaker in 1994).

John Mumper died August 8, 1863 (age eighty-two) and his wife, Jane, died January 4, 1864 (age eighty-three). They were interred in Monaghan Presbyterian Church Cemetery. Abraham Mumper died December 17, 1866 (age eighty) and his wife Mary died February 28, 1866 (age seventy-six). Both are buried in the Wolgamuth Church of the Brethren Cemetery.

Abraham Mumper's will directed that his estate be sold and the proceeds to be distributed equally among his ten children. His son-in-law Alexander Underwood purchased his farm on March 26, 1868 and built a home between Dillsburg and the mines at 45 Old York Road (owned by William Hickok in 1994).

Since these iron ore deposits were spread over a large area in cluster-like formations, we will examine them in the townships where they were located.

CARROLL TOWNSHIP

All but one (the Bender mine) of the eleven magnetic iron mines of Carroll Township were located one to two miles east of Dillsburg. The richest deposits were found along Ore Bank Road between what is now the Old York Road and Mumper Lane. The Abraham Mumper mine was located on the west side of Ore Bank Road, and the John Mumper mine was just north of it. On the opposite or east side of the road was the Smyser mine, and the Logan (Longenecker) mine was just north of it. These four mines were the most productive of the Carroll Township mines.

In 1908, a geologist estimated that 1,500,000 tons of magnetic ore had been removed in the 60-year period during which the Carroll Township mines were in operation.

Alexander Underwood built this farm on Old York Road near the iron ore mines previously owned by his father-in-law, Abraham Mumper. Today, this house just east of the borough is a private residence.

Courtesy of the Historical Society of York County

Mumper (Abraham) - Underwood Mine

The iron ore deposits discovered in 1847 were near the surface of the ground (in some cases two feet). Due to economic conditions and the fluctuating demand for iron ore, mining was sporadic; however, but when the surface deposits were depleted, Abraham Mumper was a wealthy man. After he died in 1866, his farm was sold to his son-in-law, Alexander Underwood, a Mechanicsburg attorney. Later Mr. Underwood built a home on this property and this building still exists.

JOHN MUMPER'S LEDGER

John Mumper's ledger for the period 1841 to 1864 has been donated anonymously to the Pennsylvania State Archives. Although there were many entries in the ledger, here are a few examples of ore sales.

In 1854 he sold ore to Henry Longenecker which was delivered to Mechanicsburg at $2.50 per ton. From June to November, 1,999.5 tons were sold, with October producing the greatest tonnage of 626. There were no entries from November 7, 1854 to November 29, 1855, giving evidence that the need for iron ore was sporadic. From November 29 to February 24, 1,620 tons were sold to Byron Longenecker & Co. Apparently no ore was removed from February 24 to June 4 when 124 more tons were sold to the same firm.

On February 1, 1857, 135 tons were sold to Peter Ege, owner of Carlisle Furnace at Boiling Springs. Another 227 tons were sold to Ege on April 11 for $1.12 1/2 per ton. Other small amounts were sold, some being delivered to Marietta, Lancaster County. From 1857 to the end of the ledger in 1864, there were no other entries or ore sold.

Mr. Underwood believed there had to be more ore in the area. In 1872, he hired Wrightsville Iron Company to sink shafts. Large quantities of ore were discovered at a depth of twenty-six feet, with seams up to eighteen feet thick. A tunneling operation to remove the ore was begun consisting of three drifts of 120, 180 and 280 feet. A drift is a nearly horizontal mine passageway parallel to the vein of iron ore. Each of these drifts was served by a railroad in 1885. Four vertical shafts were also sunk to a depth of 140 feet and were connected by horizontal shafts. At peak production forty tons of ore a day were extracted. No washing of this ore was required but a twelve horse power water pump using four and a half tons of anthracite coal a month was needed to pump the water from the mine. Thirty men were hired, 27 for the inside work and 3 for the outside work, each received $1.25 for a 10-hour workday.

The ore was shipped to Dillsburg, first by horse-drawn wagon at thirty cents a ton, and later by railroad cars. Ore was used at the Pennsylvania Steel and the Lochiel Works at Harrisburg, Dauphin County and the Marshall Furnace at Newport, Perry County.

NO LOCAL IRON FURNACES

It would seem natural to build and operate an iron furnace near the iron ore mining in the Dillsburg area. However there was none. The closest furnaces were at Boiling Springs, and Shiremanstown, all in Cumberland County.

Coal-fired steam engine used to power a hoist which pulled mine cars up and down the mountain. Operators Gilbert Bricker (L) and Art Hopple, Fall 1922. Gayman Collection

The Underwood was one of the most productive mines. The five large open pits were worked out by 1874 and the four deep mines were abandoned by 1887. Approximately 25,000 tons of ore had been removed from this location.

Mumper (John) - McCormick Mine

The original mine opened in 1850 and consisted of an open pit several hundred feet long. The second large open cut was about 300 feet long, followed shortly by a third cut. Although this operation adjoined the Underwood mine, a 140-foot vertical shaft failed to strike the kind of rich ore found nearby. Many smaller holes were dug. Some of these yielded considerable ore, possibly as much as 100 tons, before the mine was abandoned in 1875. The date when McCormick and Company resumed the operation of John Mumper's mine is uncertain.

McCormick and Company

McCormick and Company was heavily involved with mining in the Dillsburg area, having leases in all townships mentioned earlier. A brief history of this company reveals the importance of the iron and steel business especially during the period when iron ore was being extensively mined in the Dillsburg area.

James McCormick, a lawyer, was the founder of the company. He had two sons, James and Henry. Henry graduated from Yale in 1852 and returned home to manage his father's businesses which he did until he retired in 1897. McCormick and Co. acquired part ownership of the Paxton Furnace and Mills in 1853. By 1886 this operation was run by McCormick and Co.

The Fairview plant, a nail factory, located in present day West Fairview, Cumberland County, was purchased by McCormick and Co. in 1859. Henry McCormick was made treasurer of the plant. The output from this plant was 6,300 tons of nails and 15,000 tons of muck bars (muck bars were made from pig iron by a puddling process, which removed the carbon from the iron).

(Continued on page 8)

McCormick and Company (continued from page 7)

Most of the pig iron smelted at the Paxton Furnace, Harrisburg was made into nails at the Fairview plant. To protect its supply of iron ore, two important acquisitions were made by McCormick and Co: the Logan magnetic mine and the Dogwood hematite mine near Dillsburg.

It is interesting to note that in 1878 iron ore could be obtained from Spain and Africa at lower prices than local ore because transportation costs were cheaper and the European workmen had a lower economic living scale than Americans.

The Lochiel rolling mills and furnace were built in 1865. McCormick and Co. owned Reading Furnace iron business at what is now Robesonia, Berks County. It also had an interest in the Henry Clay and Eagle Furnace near Marietta, Lancaster County.

Henry McCormick was the father of Vance McCormick who was to become the publisher of the Harrisburg Telegraph newspaper and Annie McCormick. They summered at Rosegarden, Cumberland County on the Yellow Breeches Creek and Old U.S. Route 15.

Logan - Longnecker - McCormick Mine

John N. Logan, Dillsburg attorney, owned this mine located directly east of the two original Mumper mines. Earlier Mr. Logan sold the land which he inherited from his father for a "fair price", thinking there were no iron deposits on it. However, a mine was developed first on the Underwood property and later worked on the Fleming lease. Discovery of the rich veins of ore by Alexander Underwood made Logan consider the possibility that the same vein extended into his former property.

In 1872, buying back this land Logan paid "a very inflated price", considered by local residents an impulsive act until he found the same rich vein as Underwood's at thirty feet below the surface. This vein varied from five to fifty feet in thickness. The east drift of the shaft ran into Smyser's land who gave McCormick a lease for ten years at $1.00 royalty per ton on October 14, 1873. According to one source, G. A. Longnecker used the same shaft as the Logan mine; however, another source indicated that Longnecker leased a considerable amount of land from Messrs Underwood, Fleming and Logan.

The Panic of 1873, which lasted about four years, had a terrible effect on the nation's economy including the iron industry. However, at the end of the economic downturn the demand for iron was high and mining resumed anew in the Dillsburg area.

Twenty-one men were employed in 1876, consisting of two engineers, fourteen miners, and five outside men with wages ranging from $1.10 to $1.75 a day. An eight HP anthracite powered water pump was used almost continuously to keep the mine dry. The pump shaft was thirty-three feet deep. Two shafts were used to extract ore. Production sometimes reached as much as twenty-five tons a day.

In 1886 Longnecker sunk a shaft ninety-five feet deep (the last twenty feet was through ore) with a horizontal distance of 400 to 450 feet toward the McCormick Company operation. To the east, four additional gangways were dug toward the Logan line. In some places the ore bed was twenty to thirty feet thick.

Ore was hauled to Dillsburg for thirty-five cents a ton by private team, with loads ranging from four to six tons. McCormick and Company used the ore in its own furnaces located in Harrisburg, Dauphin County and West Fairview, Cumberland County.

It is believed this was the Logan Iron Ore Mine in 1878, located east of Dillsburg. This is the only photo found of the magnetic iron ore operations in Carroll Township.

GaymanCollection

JOHN LOGAN

The story of mining in the Dillsburg Area would not be complete without a look at the multi-faceted life of John N. Logan. He operated his farm; purchased at least one additional farm with a mine; was involved with the Dillburg and Mechanicsburg Railroad and the Harrisburg and Potomac Railroad; was an incorporator of the Dillsburg Lead, Cooper and Iron Co.; and was a director of, and counsel for, the Dillsburg Vitrified Brick and Tile Company.

Mr. Logan was born in Carroll Township on April 17, 1846. He attended Tuscarora Academy, Dickinson College and graduated from Princeton University in 1869. He farmed for several years and then studied the law. In 1873 he became the cashier of the Dillsburg Bank. For most of his life he was involved in the promotion of business in the Dillsburg area.

Ella Mae Coover became his bride on November 26, 1874. They had six children: James J. Frederick Welty, Henry, Helen Martha, Caroline E. and Eleanor. He was active in the Monaghan Presbyterian Church. He completed his law studies and practiced law in York along with his son James.

Smyser Mine

Martin Smyser's open-pit mine was directly south of the Logan property and east of Ore Bank Road from the Underwood mine. Smyser's operation was opened in 1852 by James Myers and Henry Musselman of Marietta, Lancaster County, and operated for eighteen months. Abraham Mumper purchased this mine from Smyser and worked it for six to eight months. It was sold back to Mr. Smyser, who extracted about 1,000 tons. The total amount dug from the pit was recorded as 3,000 tons. At some places the ore was twenty-five feet thick. Initially the ore was used at Marietta by Mr. Musselman but later was hauled to the Whitestown Furnace near Idaville, Adams County.

Other Mines in the Same Area

Most of the magnetic ore from the deposits east of Dillsburg was taken from the mines listed above. However, there were at least six more mines with little recorded about them. A brief synopsis of these mines follows.

"THEY HAVE NO GOOD MEN IN THIS COUNTRY"

John William Richley was born between Dillsburg and Franklintown. He had a successful Buick agency in York in a building which now houses the York County Historical Society.

Mr. Richley's father worked at the Logan mine around 1876 when the following incident occurred. A friend of his father, who had served in a Civil War company with him, was responsible for moving the carts loaded with iron ore down a long track to a turntable. There the carts could be shifted to different locations. Something happened to cause one of the ore carts to derail at the turntable and the ore spilled everywhere.

The mine at the time was run by a big, burly Englishman, William Letcher, who saw the cart spill. Without warning, he walked over to Mr. Richley's friend, pounced on him and landed quite a few punches, knocking him down several times. After the beating Mr. Letcher turned away and announced to the work crew, "They have no good men in this country".

When Mr. Richley heard this, and saw the battered and shameful condition of his friend he could not contain his seething anger and made a dive for the foreman. He hit him so hard, it sounded like a board cracking. After a few minutes of fighting, the Englishman was knocked out cold. Mr. Richley then picked him up and threw him over an embankment into a pit of water about four feet deep. He then jumped into the pit and tried to drown the foreman until his fellow workers intervened and dragged the foreman's body from the water. Mr. Letcher regained consciousness an hour later, and fired Mr. Richley.

King Mine (Also Known as the McClure Mine)

This consisted of a thirty-two foot vertical shaft with the ore vein about nine feet thick. About twenty-five tons a day were mined and about thirty gallons of water per minute had to be removed to keep it operating. It had stopped operating by 1876.

Bell Mine

Ezra Bell leased this mine to Mr. McWilliams in April 1875. The shaft was thirty-three feet deep. One source of information said the shaft was 180 feet. The ore was raised by a windlass. At least 100 tons of ore were removed. Royalty was set at fifty cents a ton for ten years. McCormick & Co. secured a lease on this mine in September, 1875.

Price Mines

There were two mines located on the Abraham Price farm. The first was worked by Henry Seidle in 1855, then leased to Rufus Castell. The mineral rights reverted to Mr. Price in 1864. Ore was found approximately fourteen feet below the surface. This large open cut exposed a vein six feet wide. A large amount of magnetic ore was removed. The land was purchased by John N. Logan in 1908.

The second was known as the Cox mine which consisted of a 300 foot slope with the vein varying from five to nine feet thick. There was a water problem which required constant pumping. In April 1887, ten to twelve men did the mining while five or six others attended boilers, hoists and other work to support the operation. At that time approximately thirty carts of ore were removed, averaging twenty-five tons a day. Mr. Cox also removed 200 tons from another location near the creek where seams were approximately six feet thick.

Grove Mine

Joseph L. Grove opened this mine in 1873. He leased the mine in 1875 to Eagle and Schultz of Newport, Perry County. Royalty was set at $1.00 per ton and 350 tons were removed. Wagons hauled the ore to Dillsburg station for fifty cents a ton where the ore was shipped to the Marshall Furnaces, Newport, Perry County. Financial difficulties forced the closing of the mine which was caved in by 1886.

Altland Mine

This operation consisted of three shafts, two north and one south, of the railway line that serviced the mines on Ore Bank and Meadow View Roads. It was located at the junction of the Bell Mine rail and the main tracks of the Dillsburg-Jauss Mine railroad. According to one source interviewed, trains ran twice a week on this line. The Altland mine was equipped to handle forty to fifty tons a day but no further estimates were given. Ore was running approximately 58% metallic iron at the fifty-seven foot level.

Jauss Mine

This mine was located at the terminus of the branch railroad, just east of Stoney Run. A seven foot vein was discovered at the seventy-two foot level. This probably was the last operating mine in Carroll Township as it was still operating in 1910.

Bender Mine

Christian Bender owned this mine, the only magnetic iron ore mine west of Dillsburg. It was located almost directly south of the area known as Beavertown, a little west of the current site of the Range End Country Club. This mining operation consisted of two small pits. Originally the Bender Mine operated for a period of four years. Stoner and Hildebrand leased the mine in 1873. The ore was black limonite mixed with clay, A six-horse team could only haul a three and one half ton load of ore a day . The going rate for hauling was two dollars per ton.

PIG IRON

The operation of a cold-blast furnace, a chemical procedure, was reasonably simple. The furnace would be filled with layers of iron ore, charcoal and limestone. Limestone was used to flux the ores. The molten ore would run down to the hearth. Several times a day the ore would be tapped into a rectangular sand mold. This mold, sometimes referred to as the "sow", would catch the cinders and slag. This same mold would also have a series of gutters, the "piglets", that would collect the clean molten iron ore. The rectangular sand mold with its series of gutters gave the appearance of a sow suckling her pig and so the name "pig iron". Before the iron cooled, the "pigs" would be separated from the "sow", which then would be broken in smaller pieces. It must be noted however, other explanations for the term "pig iron" exist.

From the time the raw ore was put into the furnace until it was tapped into the pig iron mold may have taken as long as sixty hours. Usually two tons of ore made about one ton of cast iron.

MONAGHAN TOWNSHIP

Even though the mines were located on the south side of the Yellow Breeches Creek, they were referred to as the Grantham Mines in many of the geological reports. As one drives into Monaghan Township, York County, from Grantham, Cumberland County, the mines were located less than one-half mile upstream from the bridge and immediately south of the railroad. These three mines were known as the Porter, Landis-Fuller, and Shelley. All of these mines were located on the original Knisely farm.

Porter Mine

Former Governor of Pennsylvania David Rittenhouse Porter of Harrisburg leased the land from Mrs. Mary Knisely and began mining in 1855, with royalty set at 25 cents per ton. By 1875 the excavation was 40 feet deep and was 14 feet below the water level of the Yellow Breeches creek. The vein of ore varied between three and six feet thick. Ore was removed by rail carts. The ore was first hauled to Harrisburg by wagon and later hauled to Shiremanstown. Shipments were eventually carried by the railroad.

Landis-Fuller Mine

The geological survey dated 1875 states that the Landis ore bank was better known at that time as the Fuller Mine and was located three and a half-miles northeast of Dillsburg. Daniel Landis purchased the land from Mrs. Mary Knisely. Mr. Landis opened the mine in 1863 and leased it a short time later to Mr. H. O. Shelley who operated it for about three months. In 1875 the mine was leased to Joseph Fuller of Philadelphia for twenty years. When this lease was prepared, the royalty for ore was set at eighty cents a ton with the stipulation that the royalty be paid on at least 500 tons annually. As of June of 1875, 1,400 tons had been removed, averaging about fifteen tons a day. Operating costs were figured at seventy-five cents a ton.

A tunnel was dug near the Harrisburg and Potomac Railroad into the hillside from which rail carts were pushed in and out by hand. Two drifts off of the 200-foot main tunnel headed west while the third drift was to the east. The ore was loaded into ore cars of the railroad. Ore was shipped to Wister's furnace at Hamburg, Berks County and to Daniel Ahl's furnace at Boiling Springs, Cumberland County.

Shelley Mine

H. O. Shelley mined ore in Monaghan Township from 1869 until 1878. Daniel Landis leased about eight acres to Shelley in 1874. Shelley opened a magnetic ore mine which he sold in 1875, but leased it back again in 1883. An analysis of the ore in 1874 yielded 62% pig iron in furnace tests. Another recorded analysis made in 1884 showed a 52 percent ore yield - a reasonable sampling error.

In April 1887, Mr. Shelley bought thirteen acres of the old Landis-Fuller mine. There were two shafts at that time, one eighty feet and the other forty feet. Four or five beds of ore were found separated by hard white rock-like baked slate-like sandstone.

The abondoned boarded-up remains of the Minebank Iron Ore Mine, in this 1919 photo, is located southwest of the Borough of Wellsville. **Courtesy of Mrs. Helen Snyder.**

WASHINGTON TOWNSHIP

The mining in Washington Township, referred to locally as the "Minebank", was approximately two miles southwest of Wellsville. Iron ore was first discovered around 1805, but mining by Daniel G. Altland, Jr. was not started until 1872. What was unique about this mining operation was that the land was controlled by the school trustees. The land had been donated to York County many years before the discovery of iron ore. The school house was located on the same side of Minebank Road as the mining operation. Later it was moved across the road to its present location.

The school board first leased the mining rights in 1871 to Messrs. Altland, Beitzland, and Hetrick for a period of five years. For unknown reasons the school trustees issued a new ten-year lease the following year to the same men. Daniel G. Altland, Jr. began the mining operation. A shaft was sunk thirty feet into the ground revealing a six foot vein of micaceous ore. A few months after the second lease was issued, Hetrick sold his interest to Altland. Sometime before November 1872, McCormick and Company bought Beitzland's interest. In order to gain control of the entire mining operation, McCormick agreed to pay a royalty of eighty-five cents a ton to Altland for his interest in the lease and took over operation of the mine in November 1872. The company continued to pay the school its royalty of eighty-five cents a ton. Due to the Panic of 1873, McCormick renegotiated the lease and paid sixty cents a ton to the school and fifty cents a ton to Altland.

McCormick used all the ore in its own operations. The ore was hauled to Dillsburg for $1.35 a ton, using six-horse wagons to haul the five-ton loads. By 1875 over 4000 tons had been removed. Eight men worked in two shifts for $1.35 a day. A 7HP anthracite fired water pump using two tons of coal a week kept the mine dry.

Meanwhile, Altland was involved with other explorations in the area. He stated that the mine bank vein extended southwest into Adams County for seven miles, and in the opposite direction for four miles. In spite of this claim of an eleven mile vein, apparently there were no other attempts at mining the vein.

FRANKLIN TOWNSHIP

The South Mountain Range west of Dillsburg runs in a northeast, southwest direction. Iron ore was discovered on each side of the first mountain range. Unlike the ore found east of Dillsburg, the ore deposits were non-magnetic hematite or brown ore. Three mines were located on the southeast side of the mountain. The Bender mine (mentioned above in Carroll Township) was about a mile away from the second Bender mine, which was in Franklin Township near the Dillsburg reservoir. Just a few hundred feet southwest was a large McCormick and Co. open pit mine. A second McCormick mine was about a half mile further southwest. This was in operation for only a short period of time.

On the west side of South Mountain, large deposits of ore were found in the Dogwood Hollow. Here are some additional facts about these mines:

An early photo of the Wolf Iron Ore Mine located in the Dogwood Hollow west of Tuckahoe Road. This mine later became part of the Marshall Mine operation. (Juniata Furnace & Foundry). Courtesy of Mrs. Lewis Frederick.

CHARCOAL

Until about 1837 charcoal was the only fuel which was successfully used in a cold-blast furnace. Any kind of wood was used to make charcoal as long as it was solid. Wood was readily available and was nearly sulfur free. The wood was cut in four foot lengths which varied in size from two to seven inches.

The hearth or base of the charcoal pit was a flat space thirty to forty feet in diameter and would hold from twenty-five to fifty cords. It was the collier's responsibility to construct (load) the pit following a complex and exacting procedure to reduce the risk of a "burn through".

A furnace that produced two tons of iron a day required approximately twenty cords of wood. An acre of land produced about twenty-five cords of wood. It was the practice to "clean cut" the forest, using everything but deadwood. In twenty-five to thirty years, the timber tract could be cut again.

Bender Mine

Christian Bender opened this mine in 1849. It was reported that 200 tons of ore were removed. Around 1870 a steam engine and a new type of ore washer were installed. In 1873 about eighty tons were removed from a pocket averaging about five feet thick and just seven feet under the earth's surface. The ore was hauled by wagon to the Whitestown Forge at Idaville, Adams County for smelting.

McCormick & Co. Mine

An assessment list of 1856 records a mine in operation on a farm owned by John Atticks. This mine yielded 1,500 tons of ore annually. The date of sale or lease to McCormick & Co. is unknown but the company operated it until 1868. The mine was located on the west side of Mountain Road on property currently owned by Jacob Knisley where a large pit filled with water still remains. The other McCormick mine was about one half mile to the southwest. It operated a few years and closed in 1866.

The Dogwood Hollow Mines

According to an early report, the first mine opened a few years prior to the Civil War on land owned by George Wolf. About a mile north of the Wolf property another mine was owned and operated by the Knaub family. Both of these mines were big producers of iron ore following the Civil War.

A geological survey dated 1886 included a map showing mines in the Dogwood Hollow. The mine closest to Dillsburg was named the Dogwood; then the Heck Mine; then the Brant & Arnold Mine; and last was the Markley & Shank Mine.

The next geological report on the Dogwood Hollow was twenty-three years later. This report listed the mines in the area from east to west as the Dogwood (McCormick), Knaub, Heck, Brant & Arnold (Lexier), and Markley & Shank (old Wolf). This contradicted the earlier report which indicated that Dr. Heck had purchased the Knaub Mine. The report stated that the Dogwood (McCormick) Mine and the formerly unworked area east of it was included in the Pinkerton tract. Work had started at both of these two locations shortly prior to this 1909 report.

Dogwood Mine

The first opening going west from Dillsburg was the Dogwood Mine, formerly operated by McCormick & Company. There had been extensive excavations but the sides collapsed since no work had been done there after 1884. An old shaft was found in the western end of this operation but apparently the ore had been worked-out.

Knaub-Heck Mine

In 1886 Dr. Lewis Heck of Dauphin Borough, Dauphin County purchased the Knaub Mine. It was located about 200 yards west of the Dogwood Mine. The yield was about twenty tons a day. This open pit mine was 300 feet long, 100 feet wide and eighty feet deep in places.

A pair of twenty-four foot ore washers were located near the Tuckahoe Road. Water from Dogwood Run was used to remove the silica. A tunnel about 300 feet long was dug to get into the large pit at about the sixty foot level. Ore was hauled to this level from the pit and taken through the tunnel to the washers. About one-sixth of the ore was excellent lump ore. Other cuts of twenty-five to thirty feet into the mountain revealed nothing but white clay. The work force in 1886 varied from twelve to fifteen men. The same engine that ran the ore washer also lifted the mine cars to the level of the tunnel. From this point the mine cars were taken by mule to the ore bank. Wagon owners were paid forty cents a ton to deliver the ore to Dillsburg three miles away. The ore's final destination was the Paxton Furnace, Harrisburg.

THOMAS A. EDISON VISITS LOCAL MINES

Word of the iron ore mines in Dillsburg reached Thomas A. Edison who visited them in 1903 searching for cobalt deposits. Edison wanted this mineral to perfect a storage battery. On this trip to Dillsburg he engaged a local unemployed mining expert, John Morris, to accompany him to the magnetic ore mines east of the town which were not operating at the time. Unfortunately no cobalt ore was found.

However Edison was so impressed with Morris' expertise that within a few days of leaving he sent him a telegram and invited him to report to East Orange, New Jersey for employment. During the next fifteen years Morris traveled the country in search of minerals Edison needed in his experiments. During his tenure with Edison, Morris found a deposit of cobalt ore in Tennessee in 1908. Ill health forced his retirement from Edison's employment.

Thomas Edison paid a second visit to Dillsburg on Sunday, July 30, 1905 and stayed at the Palace Hotel on South Baltimore Street. His entourage included his son Charles and his brother-in-law, John V. Miller, who was his private secretary and chief chemist. The purpose of this trip was to visit the magnetite ore mines. He was accompanied by John Morris, G. A. Longnecker, and John N. Logan. Edison was satisfied that an adequate supply of magnetite ore was available but he was not impressed with the condition of the roads, stating that the roads in his native New Jersey were much better.

In 1900 The Juniata Furnace and Foundry (also referred to as the Marshall Mine) removed iron ore in the Dogwood Hollow at the former Jacob Wolf Mine Site. To the extreme right was the ore pit where cars were brought up the incline to the building. Here the ore was washed and moved on the railway to its end where it was dumped into wagons and hauled to Dillsburg. Note mound of mud in back. The steam boiler powered the ore washer and saw mill in foreground.

Courtesy of Mrs. Lewis Frederick

MINE ACCIDENTS

Mining is a dangerous occupation but the miners and mine owners of York County were very fortunate. With more than 300 iron mines, three copper mines and about ninety limestone, clay and stone quarries, only four deaths have been recorded, all in the Dillsburg area.

William McClure lived at what is now the Jacob Knisley farm. He was killed while hauling iron ore from the McCormick and Co. mine in Franklin Township. No further details of his death are known.

Lewis Frederick related that when the Juniata Furnace & Foundry Co. bought the Heck mine, they built a narrow gauge railroad from Dillsburg to the Dogwood Hollow mine.

JF&FC asked Lewis' father, George Frederick, to manage the mine but he declined because his hearing was bad. He said that if you couldn't hear the timbers cracking giving you enough time to escape from a pending cave in, you did not belong in a mine.

Later George Frederick accepted a job keeping the tracks clear: mowing in the summer, looking for loose spikes, removing tree limbs after a storm, removing ore that fell from carts, etc. He knew the work crew would leave the mine at 5 P.M. and head for Dillsburg. When he walked home he would get off the tracks until the work crew had gone by. One day a storm came up late in the afternoon and he got off the tracks as usual to watch as the work crew went home. Then he continued to walk home. Apparently the strong wind loosened some of the empty carts which rolled down the tracks by gravity, gaining momentum as they traveled, striking and killing Mr. Frederick. Sadly, this happened because he could not hear the carts on the track.

One miner was killed in the Logan-Longnecker mine. He was run over by a mule-drawn cart that was hauling ore out of the mine.

A fourth miner was killed while standing at the bottom of a 300 foot hole. He was hit on the head by a stone about the size of a dime according to folklore.

A near disaster was averted on a May 1927 evening at the clay mine on Tuckahoe Road, Franklin Township. The mine at that time was being operated by Mineral Products Company. They had day and night crews working.

(continued on page 23)

MINE ACCIDENTS (continued from page 22)

Luke Haar was working the night crew and found the clay mine wet when he arrived for work. John Smith was working the same shift that night. While picking at the clay, tons of clay and water collapsed, pinning Mr. Haar against one side of the tunnel. Mr. Smith was working nearer the entrance and managed to flee the oncoming deluge. He jumped into one of the empty clay carts which could travel by gravity from the clay mine to the clay plant. He found George Hopple, the superintendent of the mines, at the plant. Luckily he had not gone home from the day shift. A rescue crew of workers was hastily assembled. They jumped into Mr. Hopple's auto and raced to the mine.

At the mine an attempt was made to load the debris on the carts and haul it out of the mine, but the mud was running everywhere. The rescuers could not maneuver their shovels in the restricted area of the tunnel and so resorted to using their hands as scoops. Mr. Haar was trapped about 600 feet from the main entrance on a side tunnel. Archie Bricker, who was smaller than the rest of the men, was able to slither in on top of the muck, pushing the clay-mud-water mess away as he pulled a rope back to the entombed Mr. Haar. Haar grabbed the rope offered by Mr. Bricker and slowly the crew began to pull him out of the tunnel. At one point, his hips got caught and he lost the rope. Again Mr. Bricker crawled back inside to hand him the rope. After about a two hour struggle, Mr. Haar was freed from the mine.

When Mr. Haar returned the following day for some of his posessions, he decided his mining days were over. He was offerred, and accepted, a job in the main plant where he dried clay - a much safer occupation.

The Brant & Arnold - Markley & Shank Mines (Old Wolf Bank)

The Brant & Arnold and Markley & Shank operations adjoined each other on property of McCormick and Co. Both were worked by lease arrangement.

The Brant & Arnold was about a quarter mile west of the Heck Mine and apparently was working on the same ore vein. There was no open pit. There was a tunnel located about twenty-five feet above the Tuckahoe Road. This extended over 300 feet into the mountain. A fair amount of ore was located about 150 feet from the mouth of the tunnel. They had to drill through fifty feet of barren clay until they reached a second deposit of ore-bearing clays. Different gangways were dug off the main tunnel. At the west side of the tunnel, probably about 350 feet from the mouth, an excellent supply of lump ore was exposed in a large chamber. No work was being done (1886) on these deposits since it was believed they were situated on the Markley & Shank lease. Carts in this tunnel were pushed around by hand.

The Markley & Shank lease was in the old Wolf bank. The ore was of better quality than found in the Brant & Arnold operation. This was an open pit mine which led to a seventy-five foot tunnel to reach the deeper ore deposits. The ore was four to six feet thick in the 125 foot western tunnel which had many spurs. A turntable was installed to get the carts to the various locations of ore deposits. Many times in reaching the ore layers, white clay had to be penetrated. At times 4,300 tons of ore were removed in a month but in 1886 the yield was reduced to 100 tons a month.

Marshall Mine

According to the 1909 geological report, the Marshall people operated at the mine sites of the Knaub, the Markley & Shank, the Brant & Arnold and especially the Heck mine. Marshall was also known as the Juniata Furnace and Foundry, Newport, Perry County, where they used the iron ore in their furnaces. They also provided clay to the Dillsburg Vitrified Brick and Tile Company for the manufacturing of bricks.

The Marshall Mine consisted mainly of an open pit several hundred feet long on the slope of the mountain. Ore was removed by the "milling" method. Ore was blasted from the walls of the pit. Picks and shovels were used to move the ore to the bottom of the pit where it was directed into chutes, then into cars which were hauled through the tunnel to the washers. The principal ore bearing zone was about sixty-five feet thick. Some areas of non-ore bearing clay were fifty feet thick. The brown ore was between limonite and geothite. Lying in large irregular masses, in seams, and in small pockets it was found scattered throughout the clay. Some ore textures were compact, while some were porous and cellular; others were granular, shale-like, or ocherous.

For the rest of the mining story in this area we had to rely on memories of those who worked in the mines or from those who were told stories by former miners.

<u>THE MARSHALL FURNACE/JUNIATA FURNACE & FOUNDRY CO.</u>

Iron ore from the Mumper-Underwood and Grove mines in Carroll Township was shipped to the Marshall Furnace, East Newport, Perry County. The Marshall company was involved with mining in the Dogwood Hollow, Franklin Township about 1900. George Prowell's History of York County refers to this mining operation as the Juniata Furnace & Foundry Company.

In 1907 when Dillsburg Vitrified Brick and Tile Company was formed, the company purchased clay from the Juniata Furnace & Foundry Co. holdings in Dogwood Hollow. Letters between the brick company and the foundry reveal that E. E. Marshall was vice president and his father, Alfred, president of Juniata Furnace & Foundry Co. One legal document lists Everett E. Marshall and another Edward E. Marshall with the signatures appearing to be in the same hand writing.

Juniata Furnace & Foundry Co. had an office in Philadelphia with the furnace at Newport, Perry County. Their letterhead notes mines located near Dillsburg, Pa. and their specialty was soft foundry iron. The company also manufactured stoves at Newport from this soft foundry iron.

E. E. Marshall, a Yale graduate, lived in East Newport. The Marshall family had the Marshall Furnace built in East Newport in 1872 and it operated regularly until 1900. After that time it produced rather intermittently for the next twenty years. The Marshall Furnace, a huge operation, was torn down in the 1950's.

E. E. Marshall was a staunch Republican and supposedly made the statement that if Woodrow Wilson was elected President in 1912, he would close the mines in Dillsburg. When Wilson was elected, he closed the mines.

<u>The Pinkerton Mine</u>

The Pinkerton is probably the best known of all the local mines, perhaps because of the beautiful, peaceful lake which bears the name of the former ore pit. The Pinkerton tract was an extensive open pit several hundred feet in length and a hundred feet in width northeast from the old

Dogwood or McCormick dig. Mining by the Pinkerton Construction Company of Philadelphia started in 1907. As noted before, mining had taken place earlier at this site known first as the Dogwood Mine but later as the McCormick and Company Mine.

The Pinkerton Construction Company was formed November 17, 1906 with a capitalization of $200,000, consisting of 2,000 shares at $100.00 each. The incorporators who purchased the first 200 shares were W. W. Pinkerton, St. Davids, PA, (87 shares); P. O. Pinkerton, Williamsport, PA (87 shares); and James McGinn, Philadelphia, PA (26 shares).

In 1910, a large Bucyrus steam shovel was moved from Dillsburg to the Pinkerton Mine. This all chain-drive shovel had a one yard dipper with 3/4 (270 degree) swing. It required three men to operate. One man fired the boiler, the second operated the swing and crawl, and the third operated the hoist and bucket. This shovel, which was the same kind used in digging the Panama Canal, was used to load earth and the hematite ore.

The rail line to the mine site was thirty-six inch gauge. Sections of track were laid ahead of the steam shovel to get it to the mine site. As it moved from one set of rails to the next, a second set of rails would be picked up and placed ahead of the first set. This was a tedious process.

The "dinky" engine was a narrow gauge coal fired saddle tank work horse. It hauled the ore from the Dillsburg Hollow mines to the railroad station in Dillsburg. The ore was off-loaded at the wharf where it was then shoveled by hand into gondola cars for transport to furnaces. This mine stopped operations in 1911.

Deposits of Magnetic Ore Still Remain

Possibly the final chapter of magnetic iron ore mining in the Dillsburg area was written by the United States Department of Interior in December 1947 (R.I. 4145). A team from the Bureau of Mines analyzed 125 soil samples from eighteen diamond-drilled holes (210'-555' deep) bored in Carroll Township between April 1945 to August

1945, and November 1945 to May 1946. A magnetic survey, using a Hotchkiss superdip needle, was also run between March 1945 to June 1945.

The metallurgical testing report stated, "Davis-tube tests were run on twenty composites from diamond-drilled samples. The test gave a recovery of total iron ranging from 70-97% in a concentrate assaying 60-67% of total iron". Sampling and analyses indicate the presence of other metals associated with the iron: copper, manganese, silver and gold. At least one local "panner", working Dogwood and Stoney Run has a few flakes of gold to show for a lot of effort!

Some local residents remember the team working in the area. The work crews stayed at a motel in Carlisle but lunches were provided by Ruth Webb and Mildred Schlusser at their house on what is now Meadow View Road. Mildred recalls that the crew raved over the quality of the lunches. She also washed clothes for one of the crew members. Fay Beamer's husband Bill worked part time with the crew. Several "locals" remember the mine diggings east of Meadow View Road referred to as the "Susie Hole" - no explanation!

Mrs. Arthur (Helen) Kelley was discing their farm field one early summer day (1945 or 46?) when she heard a crash above the tractor noise. At the junction of what is now Chestnut Grove and Ridge Roads, in front of their home, a car had run off the road. She and Art helped three men from the drilling crew out of the car and ditch and on to their lawn. "We washed them up and rendered what first-aid we could and after a while they left, thanking us for our help". The crew proceeded west on Ridge Road, returning to Dillsburg after drilling cores at the Smith farm.

Mineral rights and surface rights to some land between Stoney Run Road and Old York Road are reportedly still separately owned. Although iron deposits remain in the area, larger deposits of ore are found elsewhere and the use of plastics, alloys and other substitute materials - not to mention environmental standards - make the likelihood of mining for iron ore resuming in the Dillsburg area a rather remote possibility.

In 1910 immigrant workers, for the most part single men, lived in one-room shacks and tents at the Dogwood Hollow iron ore mine sites.

Gayman Collection

This is the Dogwood Hollow iron ore mines' settlement in 1910. By the early 1920s the only arifacts remaining to testify to the beehive of activity were the building foundations. Gayman Collection

OTHER IRON ORE MINES

Jeri Jones of Spring Grove prepared a mining atlas of York County. In addition to the mines mentioned above, the atlas lists other mining operations in these townships of York County, some located on old maps.

Franklin Township
- Filler's Ore Pit
- Berghart Pit
- H. A. Heiges Pit
- G. Heiges Pit
- Lemer Pit
- A. Heiges Pits

Monaghan Township
- Selby Mine
- J. Kimmel's Farm
- Elicker's Trial Shaft
- P. A. Meyer's Pit

Warrington Township
- Marshall's Pits
- Cadwalader's Pits
- Comforts Pits
- Altland's Pits
- Harman's Pits
- Gerber's Pits
- Wiley's Outcrop
- Morganthaler's Shaft
- Cooper's Pits
- Shluthower's Shaft

This 1910 photo shows the Bucyrus shovel being moved from Dillsburg to the Pinkerton Mine Site in Dogwood Hollow. It required three men to operate this steam shovel. (This was the same type equipment used in building the Panama Canal.

Courtesy of Leon Strayer

CHAPTER TWO

CLAY MINES

White Clay

Highly siliceous sericitic (silk-like containing silica) white clay was found associated with limonite in the iron mines of Dogwood Hollow and several other locations nearby, including the area known as Coffeetown. This clay was similar to that found in the Mt. Holly Springs, Cumberland County area.

The Juniata Furnace and Foundry (Marshall) was mining iron ore three and a half miles west of Dillsburg on the South Mountain when it came across the white clay. This vein of clay was reported to be seventy-five feet wide at places. The mining here was begun at the beginning of the twentieth century.

At first this white clay was considered a waste product but in 1907 the company started to sell the clay to the Dillsburg Vitreous Brick and Tile Co. for manufacturing bricks.

Later there were two other properties where clay was mined. The first known as the Russell tract was worked by the Kaolin Corporation of America, a successor to the Dillsburg Vitreous Brick and Tile Co. One source said this mine originally opened in 1916. This location was adjacent to the entrance to Lake Tuckahoe, the Boy Scout camp.

A second mine along the Dogwood Run was identified as that of a Mr. Fishel of Dillsburg. This mine was on the south side of the run and consisted of two tunnels. The clay deposit was down about thirty feet. The clay was sold to the Central Iron and Steel Co., Harrisburg and also to the Penn Tile Works Co. Aspers, Adams County.

Mr. Fishel was experimenting with the clay for manufacturing shingles, blackboards, roofing tile and other hard tile products to determine if these products could be made without firing them in a furnace. The white clay was suitable for bricks, filler in paper and paint manufacture, tile, linoleum, matches, and for refractory cement in steel mill work.

The clay mine entrance, located on Tuckahoe Road near the Boy Scout Camp, was the backdrop for the miners in this 1920s photo. First row L to R: John Hopple, Lloyd Fishel, Bruno Groupe, Hoddy Boose, John Smith, Raymond Kline, Bud Rickroads. Second row: Sonny Clepper, Charles Coxson. Third row: William Kurtz, Jake Spath, Carson Harry, Archie Bricker and Dave Smith.

Courtesy of Cumberland County Historical Society.

Black Clay

Deposits of black clay were found at the foot of South Mountain Range (at the back of Sherwood Forest development). The dates when mining operations started or terminated are not known. Edward Himes, who worked at this mine in 1927, provided the following information.

The miners worked for Dennis Asper, who leased the mineral rights. A shaft thirty-five feet deep was dug with a five foot pit to collect water. A ladder built along the side of the mine provided access for the workers. In the tunnels candles were used for light. Tunnels were dug six feet wide at the bottom tapering to five feet at the top.

Timber was used for support. Each support consisted of a mud sill at the bottom, two legs and a cap with lagging between the legs, and a heading over the top to prevent cave-ins. Trees were turned into timber supports using a cross-cut saw and an adz to square them.

A small flat car with a half barrel for a bucket ran on a narrow gauge track. The bucket was drawn to the surface with a windlass operated by two men. A cow bell was rung to alert these men that a bucket was ready for hoisting. On the surface, a narrow gauge track with a dump car was used to dump the clay at one location and the dirt at another.

Clay was mined as early as 1898 at this site in the area known as Coffeetown, currently the site of the Dillsburg Fish and Game Association. The team in the middle of the photo was owned by Willis O. Sheaffer. Courtesy of Lee Karns.

A gasoline engine pumped air into the mine and also served as a water pump when necessary. Workers labored fifty-nine hours a week at thirty-five cents per hour for a weekly paycheck of $20.65.

The black clay was hauled by truck to Aspers, Adams County to be manufactured into floor tiles by Charles and Dennis Asper.

RED CLAY

Red clay was found on the Williams farm, Dillsburg and red burning clay was discovered adjacent to the clay plant.

Workers pose on this Erie Shovel somewhere in Dogwood Hollow. Gayman Collection

CHAPTER THREE

THE CLAY COMPANIES

Dillsburg Vitrified Brick and Tile Company

Knowing that a substantial amount of kaolin clay was available from the iron ore operation at the Juniata Furnace & Foundry Co. mines in the Dogwood Hollow, a group of Dillsburg businessmen met at the Dillsburg National Bank on May 17, 1906 to form the Dillsburg Vitrified Brick and Tile Company (DVB&TC). At that meeting the following were elected as temporary officers: President, G. Bowman; Secretary, R. B. Cook; Treasurer, D. W. Beitzel. Others attending that meeting were Messrs. Ritter, Kapp, Shapley and Nesbit.

One of the first decisions was to capitalize the brick company at $100,000.

On June 14, 1906, a second meeting was held to establish the details of the incorporation. At this meeting the original subscribers were listed as:

Name	Residence
D. G. Bowman	Allen, Cumberland County
D. W. Beitzel	Dillsburg, York County
R. B. Cook	Harrisburg, Pa.
H. E. Ritter	Harrisburg, Pa.
John N. Logan	Dillsburg, York County

Tuesday afternoon, July 10, 1906, some of the investors met in the office of Edward W. Cooch, Esq. in Wilmington, Delaware and decided to incorporate in the State of Delaware. The stated purpose of the corporation was "to manufacture bricks and tiles, and all kindred products from clays and other siliceous materials, and earthy substances, and for that purpose to buy, lease, own, contract for and sell land and other real estate, and to buy, sell and deal in all kinds of manufactured bricks, tiles, clays, kaolins and other earthy and volatile substances, and all necessary machinery for conducting and carrying on the same".

At a meeting held on July 19, 1906, permanent officers were elected for one year. The officers were:

D. G. Bowman - President G. W. Cook - Secretary
H. W. Stubbs - Vice President D. W. Beitzel - Treasurer

The brick company decided on July 26 to contact Everett Marshall, Vice President of the Juniata Furnace & Foundry Co., about securing land, leases and right of way.

When the brick company officials met in August at the American House in Mechanicsburg, they decided to contact the Cumberland Valley Railroad about laying a spur track from the main line in Dillsburg to the plant site. Also at this time, Thomas Hampton of West Elizabeth, Pennsylvania was asked to submit a proposal for construction of a brick plant.

The brick company agreed to a contract with E. E. Marshall, Juniata Furnace and Foundry, at its August 24 meeting. At that meeting R. B. Cook resigned as a director to make a place on the board of directors for E. E. Marshall of Newport. At the next meeting, Mr. Marshall was elected as Vice President. Through the agreement, the brick company and the mining company co-owned 50.8 perches of land needed for the railroad right of way to the plant about one half mile west of Dillsburg.

At a special meeting on September 27, the stockholders were listed as D. G. Bowman, D. W. Beitzel, H. E. Ritter, G. W. Cook, R. B. Cook, J. N. Logan, E. W. Cooch, and H. Grub.

On October 15, 1906 the company decided to send about ten barrels of clay to the American Clay Machinery Company in Bucyrus, Ohio for the purpose of conducting tests on the clay. It was reported at the next meeting that twelve barrels of five different clays had been shipped. Meanwhile a committee was reviewing the proposed plant site. On December 12 a contract was given to Thomas Hampton of West Elizabeth, Pennsylvania to build the clay plant and a committee was appointed to secure terms with the Cumberland Valley Railroad to build the spur with an estimated cost of $5,980.

The DVB&TC began having financial problems on May 13, 1907 when work stopped on plant construction because the contractor failed to complete his contractual agreements with the brick company. The company notified the surety bond company and rescinded Mr. Hampton's contract as well as his position as general manager. The board appointed a committee to work on completion of the plant. Under this arrangement the building was soon completed and the board decided to erect an office building in June. The brick office building was twenty feet by thirty-six feet with ten foot ceilings. Charles L. Blissard was hired as the new general manager.

Machinery for the plant was purchased from American Clay Working Machinery of Bucyrus, Ohio. On October 30, the brick company voted to have water installed by the Dillsburg Water Company.

The board met on November 19, 1907 and discussed the possibility of discontinuing the operation of the plant because the economy was suffering an economic downturn (Panic of 1907). But the decision was made to continue as rapidly as possible. At its December 6 meeting the board authorized its first purchase of clay from the Juniata Furnace & Foundry Co.(JF&FC) for $392.

The cost of clay from JF&F was $1.00 per ton with an additional fifteen cents per ton charged for delivery.

The company president was authorized to borrow funds to cover the payroll with each director being responsible for his pro rata share of December 12. Two weeks later the treasurer resigned. On January 31, 1908, the general manager was asked to sound out the workmen at the plant regarding a cut in their wages and if agreed upon, he was to put this reduction into effect.

The general manager reported at the February 14, 1908 board meeting that ninety-five bricks were ready to burn. His personal preference was walnut coal but reported the following week that it was not performing well and decided to use coke. A second payment for clay was authorized to be paid to JF&FC for $386. In the next few weeks 28,000 bricks were made.

The brick company could not reach maximum capacity of brick production due to moisture content of the Jersey clay which was mixed with the local clay. This problem was overcome by building a drying shed. Meanwhile the financial condition worsened and the president was authorized to make a loan from the Dillsburg National Bank with each director assuming his share of financial responsibility. Also notes of $1000 each for sixty, ninety and 120 days were issued to the American Clay Machinery Company since payment for the plant equipment was overdue.

Discoloration of bricks was a problem and it was soon discovered that oil in the clay caused the discoloration. Somehow a solution to this problem was found and the plant began making 16,800 bricks a day.

By April 1908 the balance in the treasury was $83.30 and each director was assessed $100. By May the account was slightly overdrawn and directors again were requested to pay another $100. A brick agent from New York was invited to the plant for an inspection to determine whether he would be interested in representing the company.

The financial condition of the brick company was discussed at the stockholders meeting on July 3, 1908, and it was agreed to secure a second mortgage for $80,000. The following officers were elected for the coming year: G. W. Cook, President; E. E. Marshall, vice-president; D. G. Bowman, secretary; and Frank R. Peckman as assistant secretary and treasurer.

Typical of one of the many surviving Dillsburg bricks. Photo by Jacob Heisey

In August, two carloads of bricks were sold in Camp Hill and a carload was sold to A. B. Harnish in Mechanicsburg. By October the plant was selling many bricks, including $2,500 worth to JF&FC.

G. W. COOK, Pres. Contracts accepted subject to delays caused by strikes or other causes beyond our control. R. B. COOK, Sec'y.
E. E. MARSHALL, Vice-Pres. C. L. BLISSARD, Gen. Mgr. D. W. BEITZEL, Treas.

Dillsburg Vitrified Brick and Tile Company

MANUFACTURERS OF

HIGH GRADE FACE BRICK

IN WHITE AND LIGHT SHADES

DILLSBURG, PA.,

An Invoice that was never used is now a part of Bob Gayman's Collection

The brick company was having trouble securing the $80,000 second mortgage, and directors were periodically asked to contribute $50 or $100 at each board meeting. The directors agreed to offer the plant to Mr. Bass for $145,000 and royalty on clay was to be seven and one half cents per ton for the first 500,000 tons and five cents per ton thereafter. This offer was rejected.

At the May 22, 1909 board meeting George W. Cook was appointed to negotiate a lease of clay lands with Huron Products Co., Philadelphia.

A special meeting of the stockholders was held on June 1, 1909. The purpose of the meeting was to consider plans for reorganizing the company or to plan for the sale of the entire property. A committee of George W. Cook and John N. Logan was directed to either pursue reorganization or to secure a buyer. Each director was asked for $750 to pay off the note at City Bank of York and provide money for the expenses of the committee mentioned above.

By September everything was coming apart. General Manager Blissard resigned to take a new position in Wisconsin. A special meeting of the board was held at the Commonwealth Hotel in Harrisburg to consider the results of a meeting of bondholders and creditors. Since the company was unable to pay the interest on the October 1, 1906 bond issue secured by a mortgage, the company waived the demand in writing for a period of three months and agreed that the bondholders may make written request to said trustee to sell the property. Eighteen of the twenty four bondholders were present. The plant had been idle for some time and the resources of the company were absolutely exhausted. It was determined that the interests of all parties would be best served by a foreclosure. J. W. Wetzel, Esquire of the Merchants National Bank, Carlisle was selected as a trustee to buy the property on foreclosure and organize a new company if possible. It was stated that the failure of the present company was caused by spending too much money on the erection of the plant and leaving too little for its operation, a cash flow problem.

On November 20, 1909 a board meeting was held at the Hotel Central in Dillsburg with a view of re-organizing. A public auction of the property and plant was scheduled for 1:00 p.m. that day but no action was reported in the minutes.

The last entries in the brick company's official minutes were dated February 26, March 19 and May 7, 1910, with no definite action taken at any of these meetings. The correspondence of the brick company contains a vast amount of letters between bankruptcy trustee J. W. Wetzel and the brick company attorney, John N. Logan, regarding their efforts to secure a buyer.

The following story in the March 22, 1911 edition of the Harrisburg Telegraph newspaper covered the the final chapter of the Dillsburg Vitrified Brick & Tile Company.

"The Dillsburg Vitrified Brick and Tile Company's plant was sold Tuesday afternoon to A. U. Shauer, of Lewisburg, Pa. for $1,500. This plant, which has a daily capacity of 20,000 brick, was built three years ago at a cost of $150,000. It has been idle for about two years. The new owner expects to have the plant in operation within two weeks. When in full blast it will give employment to forty men."

Dillsburg Clay Products Company

What happened after the sale of the plant to Mr. Shaurer is unknown, but a letter in the DVB&TC records state the brick plant was sold at public sale on August 4, 1911 to John Hosfeld, a builder and contractor of Shippensburg, PA for $5,400.

The Dillsburg Clay Products Co.(DCPC) was formed on September 26, 1911 with John Hosfeld as the major stockholder with eighty shares. Four other investors each owned five shares including John W. Wetzel who was listed as the company secretary.

Sometime during the next eight or nine years the old round brick kilns were removed and replaced with long brick ovens to create better bricks. Apparently this was a bad business decision which may have caused the company to fail.

Penn Products

About 1920 Penn Products (PP) took over the operation of DCPC and added to the clay plant structure. According to a newspaper article brick making ceased in 1922.

The clay was brought in from the mine and washed to remove the sand and other impurities. Indigo dye was added to bring out the whiteness of the clay. Water was then pressed from the clay, leaving large cakes. These cakes were baked in large ovens. The baked cakes were ground into a powder, just like flour, and placed in bags weighing 100 pounds. The clay was then shipped by railroad cars to be used in paints, talcum powders, candy, cosmetics, and also linoleum by Armstrong Company, Lancaster County.

Mineral Products Company

About 1927, the PP's operation was called the Mineral Products Company but no incorporation records appear at the Pennsylvania Secretary of State's office.

It was during the operation of the Mineral Products Company that the cave-in trapping Luke Haar occurred. Sometime during World War II (1941-1945) rails were removed from the dinky track to aid the war effort. Fortunately some were rescued and are the property of NYCHAPS.

Dillsburg Kaolin Company

A Pennsylvania Department of Internal Affairs report on white clays of Pennsylvania, dated July 1934, stated that the clay mine on the Russell property was last worked by the Kaolin Corporation of America but has been idle for many years. A letterhead exists with the name Dillsburg Kaolin Company of 1328 Chestnut Street, Philadelphia with the notation of "Mines and Works, Dillsburg, Pa", but the company was not found to be incorporated with the state. No operating dates have been established.

Mines and Works, Dillsburg, Pa.

DILLSBURG KAOLIN COMPANY

GEO. W. EMERY, President
A. S. WOOLLEY, Vice-President
T. W. McNEIL, Secretary-Treasurer
E. L. FEEMAN, General Manager

General Offices
615-618 HALE BUILDING, 1328 CHESTNUT STREET

PHILADELPHIA

This letterhead was found and donated to Bob Gayman.

Z-W Manufacturing and Products Corporation

From the summer of 1944 to the early 1950's the clay was removed by Z-W Manufacturing and Product Corporation of Columbia, Lancaster County. The processed clay was shipped to Baltimore to serve as a base for herbicides.

A young soldier returning home from the war remembers visiting the plant and seeing both white and black clay being processed. The processed clay was sold to Central Chemical, Baltimore, MD.

<u>Mauna Mining Company</u>

The last occupant of the clay plant was the Mauna Mining Company which ground soapstone in the 1960's. The company brought the soapstone from Idaville to be used for talcum powder and cosmetics. No incorporation records were found.

<u>Remembrances of a Clay Mine Worker</u>

Gilbert Bricker was the brother of Archie Bricker who helped save Luke Haar's life at the same mine in 1927. Gilbert worked at the clay mine from March 15, 1922 to March 15, 1923. It was easy to remember the date since it was his birthday. During that work period he boarded at the home of George Hopple who was the superintendent of the clay plant and the clay mine. Mr. Hopple came here from Mt. Holly Springs, Cumberland County, where he worked for the Allen Clay Co. which operated a mine at Toland, Cumberland County. His son, Art, worked at the clay plant and his other son, John, worked at the clay mine.

When George Hopple came to town, he along with Richard "Old Man" Russell (who lived on Cold Springs Road) opened the clay mine at the entrance to what is now Camp Tuckahoe. Since Mr. Hopple had mining experience, he moved ahead without Mr. Russell's instruction. This caused friction between the two since Mr. Russell wanted to be in charge. So he started carrying a gun. One day "Old Man" Russell mysteriously disappeared and it was said later that he went down south somewhere.

Mr. Bricker remembered riding to work in the Dogwood Hollow on the dinky train and stopping along the way to pick up other mine workers as they proceeded to the mine from the clay plant. As the men would ride to work, they would pass the closed Marshall mines, and only foundations remained where the miners' homes once stood.

George Fissel was in charge of the mine at that time. Mr. Bricker's job was to operate the shovel which was used to strip off the overburden of the mine to get to the clay. Previous shovel operators at the clay mine were Richard Russell and Lewis Frederick. Then about three foot "monkey drift" round holes were dug from inside the mine to the surface. A mine car (which was smaller than the dinky cars used to haul the clay to the plant for processing) was placed under these holes. The loose clay was dumped down the monkey drifts until the mine cart was full. The cart was moved by mule from the mine and dumped into a bin. The dinky railroad cars were in turn filled by gravity for the trip to the clay plant.

Ira Knaub worked with the mules. Carson "Carse" Harry, along with two helpers, was in charge of shoring up the tunnel with six to ten inch timbers as they worked their way into the mine. Mr. Bricker further recollected that about 100 feet into the tunnel of the clay mine, iron ore deposits were breached which were about 60% iron, one of the best veins around but not worth enough in 1922 to remove. At this time twelve people worked at the mine including Paul "Bucky" Frederick, brother of Lewis Frederick.

Workers take a break at Dillsburg Clay Plant around 1920. Standing L/R George Hopple, John Mummert, Charles Williams, Calvin Albert and Russell Wolfe. On shovel is Johnny Hoover (operator) and Gordon Cullings. Courtesy of Mrs. Eugene Kimmel.

The Dillsburg Clay Plant, on Mountain Road about one mile west of the Borough, was built in 1907 by the Dillsburg Vitrified Brick and Tile Company. It was operated sporadically until it was demolished in the 1960s. The brick building at the left served as the company office and is currently a private residence. Gayman Collection.

The Kuntz Limestone Quarry near Beavertown at the time when horses and wagons were utilized to haul limestone. Courtesy of Leon Strayer

CHAPTER FOUR

LIMESTONE QUARRIES

Limestone formations and outcroppings appear at various areas in northern York County just as they appear in abundance in adjoining Cumberland County. Although mining iron ore was either tunneling or open pit, quarrying for building stone, slate or limestone was open excavation.

Kuntz's Quarry

John Kuntz (one of the principles involved in building the church in Beavertown and elected its local minister in 1881) lived on a fifty acre farm at the western edge of Beavertown on Cold Springs Road, west of Dillsburg. The quarry was located on this farm. The limestone was a coarse conglomerate similar to Potomac Marble. Again, the same fine quality found in Cumberland County was found here. There were four lime kilns at the quarry. These converted the quarried limestone into lime. In 1875, a geologist reported that Kuntz sold his lime at the quarry, mostly to his neighbors, for nine cents a bushel. He employed eighteen men and was in the lime burning business for twenty-three years. Red and yellow ocher used in making paint were also found at this Beavertown operation.

When the NYCHAPS Mine Committee visited this quarry in the early 1990's, a good-sized stream flowed into a hollow at the bottom. It then disappeared as opposed to filling up the hole and forming a pond as seen in most of the iron ore mines.

LIME PURCHASE UNPAID

Lewis Nelson and his son, Lloyd, "burned" lime at the Kuntz limestone quarry. It was then sold to farmers. One farmer, "Mercy" Leer, got some lime from Lewis Nelson but did not pay for it. Mercy saw Mr. Nelson sometime later, stopped and said, "Mr. Nelson, I guess you'd like to be paid for your lime?". Mr. Nelson replied, "Yes, I can use the money". Mercy said, "I got it, it's in the 'Dillstown' Bank, it ain't doin me no good nor you no good." "Giddy up.", he said to his horse, rode away and never did pay for the lime.

The Kuntz Limestone Quarry had rails and carts installed to remove the limestone, replacing horse and wagon teams shown earlier.

Courtesy of Leon Strayer

The Kuntz Quarry lime kilns were located west of Beavertown. Here limestone was "burned" to provide lime for farmers' fields.

Courtesy of Leon Strayer.

The work crew posed for a photo at the Kuntz Limestone Quarry kilns where the "burned" or finished product is removed.

Courtesy of Leon Strayer.

At the Kuntz Limestone Quarry off Cold Springs Road, Beavertown, Lloyd E. Nelson holds the sledge hammer, while his father, Lewis Nelson, (in Derby hat) holds the rod. The third man is unidentified. Courtesy of Mrs. Robert Byers

Williams Grove Quarry

The Williams Grove quarry was in the extreme northwestern corner of York County, lying east of Dogwood Run and a few hundred feet from where Dogwood Run empties into The Yellow Breeches Creek. This operation began in 1898.

In 1930, the working face was approximately 500 feet long, varying from eighty-five to 100 feet high. There was a heavy clay overburden. From time to time small caves were exposed as the layers of limestone were excavated. These caves had characteristic stalagmites and stalactites.

For the first nine years it provided ballast for the Pennsylvania Railroad. Later, most of the crushed limestone was used for building state highways especially when concrete highways began replacing dirt or oil and chip roads. This quarry yielded a fine grade of crushed stone of dense, hard blue dolomite.

This quarry was operated for years by Strayer and Lefever Brothers of Dillsburg. It ceased operating after nearly all the limestone had been removed.

Other Smaller Limestone Quarries

Limestone was found in a variety of areas, not unlike the deposits of iron ore. Some of these quarry sites were identified in geological reports but little detail given.

Welty's quarry was listed as existing one-third of a mile north of the Dillsburg railroad depot of the Dillsburg and Mechanicsburg Railroad. Some of this limestone was greenish gray while some was white in color.

A thin, not too pure, limestone conglomerate was quarried at the roadside one-half mile southwest of Dillsburg (this was along Golf Course Road in Carroll Township).

The work crew of the Williams Grove Limestone Quarry pose with their tools. The third person from the left is Lloyd Nelson, the other men can not be identified.
Courtesy of Mrs. Robert Byers.

Another limestone quarry was situated about one-half mile northwest of the area known as Uniontown (Mountain and Union Church Roads).

Lime kilns were usually built near the limestone source. A kiln was located near Range End Country Club and another on Mountain Road about 200 yards south of Tam Systems, Inc. This kiln near Tam Systems probably processed the limestone from the previously mentioned Uniontown quarry. Without a doubt many more kilns existed in the area where local farmers burned limestone for use on their own farms.

DINOSAURS IN THE AREA?

In view of the current fascination with dinosaurs, it is of interest that while quarrying sandstone in the 1930's at the Trostle Quarry near York Springs, Adams County, dinosaur footprints were discovered. They were preserved in Triassic sandstone. This same baked Triassic sandstone was found in the local magnetic mines. Since York Springs is a mere nine miles to the southwest of Dillsburg, dinosaurs probably roamed here.

Lime was processed by placing limestone in these kilns which were usually built into the side of a hill, with an opening near the floor for removal of the finished product. Another opening at the top allowed the gases created during processing to escape. Wood was used for fuel and was placed in the lower one-third of the kiln and then limestone chunks were placed on top of the wood through a large opening at the top. The burning process was begun with the ultimate goal of obtaining "slacked" lime in the form of a fine powder. This lime powder was spread on farmers' fields to increase crop production.

CHAPTER FIVE

COPPER MINES

Three copper mines are recorded in a newspaper story of the mines of York County. None of these were in the Dillsburg area. However, local oral accounts indicate that there were three copper mines in the Dillsburg area but nothing has been written about them.

The first was located on the Adams/York County border, just off County Line Road and Old U.S. Route 15. This Franklin Township mine was worked for a short period of time. When the cost of mining the product exceeded the market value of the copper, the operation was discontinued and the mine was abandoned.

The second was in Barren's Valley on Cabin Hollow Road near the intersection of Carroll, Warrington, Washington and Franklin Townships. There were no tailings near the entrance of the mine when visited in the 1990's, possibly the tailings were used to refill the old mine hole.

The third was in the Rossville area northeast of the Old York Road, not too far from Squire Gratz Road in Warrington Township.

All the mining operations probably were active around the 1880's.

John Kuntz, who owned the limestone quarry, was one of the organizers of the Dillsburg Copper, Lead and Iron Company. He was interested in the copper business and traveled to New Jersey, New York, West Virginia, Virginia, Ohio, Indiana, Illinois, Iowa, Minnesota, Wisconsin and Michigan to inspect copper ores.

When the Dillsburg Copper, Lead and Iron Company was formed October 11, 1883, its stated purpose was mining and smelting copper, lead, and iron ore. Some copper was smelted. Apparently the locally mined copper ore was hauled to this smelter which was reportedly located along the Dillsburg and Mechanicsburg Railroad close to Mount

Olivet Campground, Carroll Township. No artifacts were identified in 1993 to confirm this location. It is unknown how long this smelting plant was in operation.

The incorporators of the Dillsburg Copper, Lead and Iron Company were:

NAME	SHARES	RESIDENCE
Joseph Deardorff	750	Dillsburg, York County
John N. Logan	550	Dillsburg, York County
John Kuntz	400	Dillsburg, York County
Lewis Kuntz	400	Dillsburg, York County
William Raker	300	Sunbury, Northumberland County
A. R. Hurst	450	Mechanicsburg, Cumberland Co.
William Raber	350	Treverton, Northumberland County
H. H. Mercer	300	Mechanicsburg, Cumberland Co.
U. H. Vanarsdale	550	Mechanicsburg, Cumberland Co.

Copper was also mined in Hunterstown, Adams County. The copper ore was hauled to Dillsburg by the Galloway Brothers. In 1885, a prospector, Paris Erb of Hunterstown, excavated some yellow rocks and shipped it to the Dillsburg refinery as "Copper Gold". It did contain some gold but was mostly copper. However, in less then two years he closed his mine. The mine was then reopened in 1903. Workers reported gold streaks in the copper ore but not enough gold was extracted to recover the cost of shipping it to refineries. The mine operated sporadically from 1908 until 1916.

Copper deposits were found in 1951 when excavations for the first high school of the Northern York County School District were made. The school is built south of Dillsburg on the west side of PA Route 74 in Carroll Township. It is said this copper vein runs in an easterly direction toward Stoney Run Road

Maud, the mine mule, poses with clay mine workers while waiting for the "Dinky" engine to take them to the clay plant. (This is the same crew that appear in photo on page 32)
Courtesy of Mabel Bricker Nell

CAPITAL $100.000.

No. 414

Shares

The Dillsburg Copper, Lead and Iron Company.

SHARES $20 EACH.

SHARES $20 EACH.

THIS CERTIFIES that A. R. Hursh is entitled to One Shares in the Capital Stock of The Dillsburg Copper, Lead and Iron Company, Transferable only on the books of the said Company in Person or by Attorney on the surrender of this Certificate.

WITNESS the seal of the Company and the signatures of the President and of the Secretary at Dillsburg, Pa., this Sixth day of March A.D. 1885

This one share of stock of the Dillsburg Copper, Lead and Iron Company, worth $20.00, was issued to A. H. Hursh on March 7, 1885 but was never delivered.

Courtesy of Jay Lininger

CHAPTER SIX

TRANSPORTATION

When ore was first mined in the Dillsburg area, it was transported by wagon and hauled directly to the smelting furnaces. The wagons were usually pulled by a team of two, four or six mules or horses. Some wagons carried two and one-half to three tons while others could carry as much as five tons. Drovers were paid by the ton of ore hauled and rates over the years varied from thirty cents to $1.35 per ton. After the Dillsburg railhead, which consisted of a siding and wharf, was completed, the wagoners hauled the ore to the railhead.

The furnaces were far removed from the mines, so only a limited amount of ore could be transported each day by wagon. With so many mines in operation and so many teams on the road, something had to be done to improve ore delivery to the furnaces.

Railroads were being expanded all around, so it was only natural for the local mine owners as well as the Dillsburg merchants and businessmen to be anxious to initiate a railroad service for the developing town. Shiremanstown, too, was growing and interested in rail service. Pressure by the Dillsburg interests spurred the formation of the Dillsburg and Mechanicsburg Railroad Company (D&M) on November 1, 1871 as part of the Cumberland Valley Railroad System. The length of the railroad from its intersection with the mainline (near Mulberry Road west of Mechanicsburg, Cumberland County) to Dillsburg was 7.7 miles.

Thomas W. Paxton, Carlisle, Cumberland County, was the contractor chosen to build the line. It was completed in October 1872; however, it did not open until 1873. The mines continued to rely on the wagons to get the ore to the Dillsburg wharf. In 1888, a railroad spur was built from Dillsburg to the Bell, Logan and McCormick Mines

This wagon pulled by four mules was the type of wagon that hauled ore from the mines. In this picture the wagon was being used to haul stones to repair streets in Dillsburg.

Gayman Collection

located in Carroll Township. Later, this spur was extended east to service the Jauss Mine. The railroad bed was still evident in 1992.

Around 1901, the Juniata Furnace & Foundry Company (Marshall) was mining in the Dogwood Hollow. In 1902, the company built a narrow gauge railroad from Dillsburg to its mining operation to haul the iron ore and later clay. When the Dillsburg Vitrified Brick and Tile Company had a standard gauge spur built to its plant from Dillsburg (1907), it was noted in a letter that this spur paralleled the narrow gauge railroad of the Juniata Furnace & Foundry Co.(JF&FC) The brick company bought their clay from the JF&FC.

The Pinkerton Mine apparently used the narrow gauge railroad of the JF&FC for hauling its ore when it operated from 1907 to 1911.

Passenger service was initiated on the D&M Railroad in 1873. The service consisted of one train, each way, once per day. The trip to Mechanicsburg took forty-five minutes. In 1881, there were two trains operating in both directions and by 1895 this increased to four daily trains. Passenger service was again increased to eight trains a day in 1910. This schedule was curtailed during World War I to six daily trains. The D&M was unique in the Cumberland Valley Railroad system because it was electrified in 1905, providing local trolley service. The last passenger train to Dillsburg ran on January 15, 1928.

When the Cumberland Valley Railroad was merged into the Pennsylvania Railroad on June 2, 1919, the D&M was included in the merger. The last freight train to Dillsburg was in 1973 and the track and ties were removed in 1980.

The mines in the Dogwood Hollow were served by a network of narrow gauge (36") rail lines. Small 0-4-0 saddletank engines were used by several of the mining company to pull the ore cars from the mines to the washers. Later a line carried clay to the brick and clay plant at Beavertown. Gayman Collection

SECTION II

REGIONAL GEOLOGY

by

Jay L. Lininger

INTRODUCTION

The founding of the Commonwealth of Pennsylvania encompasses a story of rich and varied history. Although the repositories and archives of the state contain abundant reference material, they are surprisingly devoid of detailed information concerning the establishment and early history of Dillsburg and the surrounding vicinity. "MINES OF THE DILLSBURG, PENNSYLVANIA AREA' is an historical perspective which attempts to construct a detailed account of one facet of Dillsburg history which has been grossly understated in the literature. It is a tapestry woven from many partial and obscure references which by themselves present an incomplete and often confusing picture. The information contained in this monograph provides the reader with a more complete picture of a rapidly vanishing aspect of Dillsburg's heritage, and has preserved it for the enjoyment of future generations.

The earliest settlers of the Dillsburg area were typical of most pioneers that helped colonize the Commonwealth of Pennsylvania. These early citizens, generally of northern European ancestry, were seeking to establish roots in the new colonies. They brought with them great skills in homesteading and development of the land. Agriculture, the first choice of sustenance, was the precursor of small rural communities which dotted the fertile countryside. These small agricultural communities became the backbone of the early development of the nation. Dillsburg was one such community. The first settlers to arrive in the vicinity would certainly have been attracted to the fertile land and meandering streams so perfectly sheltered by the South Mountain Range located to the north and west. Here were all the requirements for ideal farming, bounded by an unlimited supply of hardwood timber. A handful of eighteenth century settlers chose to go no further than this idyllic setting and proceeded to clear the land and establish the farms that provided the economic foundation for the new community.

Dillsburg was established in 1808 on 504 acres of land owned by it's namesake, Matthew Dill. The village flourished and soon attracted the other types of businesses and forms of commerce required to support an agricultural community. On April 9, 1833, the settlement was officially incorporated as a borough and was able to boast that it contained 40 dwellings and a population of 200 residents. For the next century and a half, the Dillsburg area retained its bucolic setting and agricultural framework, seemingly untouched by all but isolated events. By the 1970s an influx of development had begun to change the face of the community. Spurred on primarily because of it's proximity to the increasingly busy corridor of U.S.Route 15, Dillsburg began to lose its isolated, small town image. Along with an increase in the population of the area, came an increase in interest in the area's history. When the Northern York County Historical and Preservation Society (NYCHAPS) was founded in 1984, it fulfilled a great need in its mission to preserve the historical heritage of the area.

NYCHAPS is a vibrant and growing organization which has made important strides in preserving vast amounts of regional history in a few short years. Its support of this publishing effort was vital to the success of saving information on one of Dillsburg's lesser known areas of commerce. The early settlers who had a great appreciation and kinship of the area's fertile soil, also had an appreciation for the mineral wealth found in the soil and surrounding hills. Through several centuries of the region's history, efforts were made to exploit the natural resources found in the area.

Some attempts at mining were quite successful while others were failures. Unfortunately for those interested in local history, these mining entrepreneurs saw little value in preserving a record of their activities for posterity. Thus a committee of interested NYCHAPS members, led by Bob Gayman, assembled for the purpose of seeking and organizing all available information relating to the mining, quarrying and processing of minerals in the Dillsburg area. The task also included an attempt to locate and make on-site visits to mining and quarrying locations in the region. Attempts were made to locate all available pictures and preserve them through high quality photo reproduction. Oral interviews with local citizens who had past association with, or knowledge of, local mining operations were conducted and recorded. All reasonable and logical sources of historical information and material, including the Pennsylvania State Archives, were consulted and utilized. Efforts were also made to acquire and preserve local mining artifacts. The findings of this research are presented in this publication of the NYCHAPS Mines Committee. Although there are unreconciled gaps in the chronological history of the story, the report is the most complete history of Dillsburg area mining assembled to date.

CHAPTER ONE

GEOLOGICAL FRAMEWORK

In order to better understand the mining history of the region, the reader is provided here with basic information regarding the geological processes which brought about the formation of these valuable mineral deposits. For those interested in learning more about the geology of the Dillsburg area, a list of selected readings is provided.

The science of geology covers a vast spectrum of the development of earth and space. In order to distill the information needed for the purposes of this report, it is necessary to provide guidelines in three distinct geological disciplines, topography, age, and rock types. These three general disciplines will be used to describe the economic mineral occurrences located in the Dillsburg area.

1. TOPOGRAPHY: Throughout the history of mankind, there have been borders. Borders are established in different ways, either by geographic limits, such as rivers or mountains, or by geopolitical means, such as war or negotiations. The science of geology establishes borders by geographical limits within landforms. These are referred to in the geological textbooks as provinces. York, Adams and Cumberland Counties embrace parts of the Blue Ridge and Piedmont provinces of the Appalachian Highlands. On a more regional basis the geographic landforms can be further refined and defined. The Dillsburg area, geologically described, is based within landforms known as the Triassic Upland, the South Mountain and Cumberland Valley (see the accompanying illustration on page 68)

These landforms are visually evident and easy to discern by an observer. Commercially valuable mineral occurrences located within the scope of this report are to be found in the Triassic Upland and South Mountain geological regions.

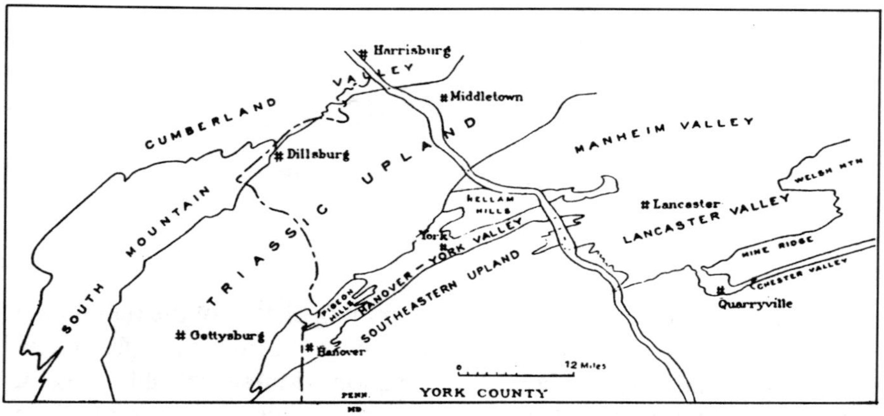

Geologic Provices (or subdivision) of the Dillsburg area as adapted from the "Geology and Mineral Resources of York County, Pennsylvania; PA Geological Bulletin C-67, Stose and Jonas, 1939.

2. FORMATION AGE: The study of rocks and the understanding of their formation is closely connected with the age of the formation. Determination of the age of rock formations has been one of the complex problems which has taken many years to understand and classify. It is often difficult for the layman to comprehend that two different rock formations, which contact or touch each other, are often millions of years older or younger that the other. This paradox is possible because the geological processes which bring about the formations of the rocks are non-ending and often violent in nature. Our understanding of these processes is always being fine tuned and further enhanced by new information or new interpretation. In recent years, the science of geology has embraced the concept of plate tectonics, which states that the earth's crust is made up of a family of land masses, scientifically known as plates, which are constantly in motion, floating on a sea of molten rock. Such movement of land masses in our prehistory of the region could account for an uplifted land mass such as the nearby South Mountain.

Geologists estimate the age of the earth at four and one half billion years. This span of time has been subdivided into three eras (Paleozoic, Mesozoic and Cenozoic). These three major eras have been further subdivided into periods, whose dividing lines have been determined by major events within the movement of the plates. For comparison, one could visualize a yearly calendar. A calendar year could be compared to the entire geologic age, with the eras represented by seasons and the periods represented by months. The most significant differences would be in the length of the periods. Unlike the months in the analogy, however, they are not equal in length, and widely vary.

An illustration of the geological age calendar is provided to further clarify the age sequence for the reader. On the following chart, the reader will note that the order of the age of rocks in the Dillsburg area is as follows: (A) South Mountain landform; Precambrian and Cambrian Period 500 to 800 million years in age. (B) Cumberland Valley landform; Ordovician Period, 430 to 480 million years of age. (C) Triassic Upland landform; Triassic Period, 180 to 225 million years of age.

The area occupied by the Borough of Dillsburg is significantly newer in geological age than the South Mountain Range which lies directly to the west and northwest of the town. Beneath the Triassic Upland however, are basal rocks upon which the Triassic rocks formed. Although these basal rocks are not seen or known, they are likely to correlate to the older Precambrian and Cambrian rocks of the South Mountain. A geological event taking place during one of the tectonic plate movements induced a tearing apart and uplifting of the rock mass we now refer to as the South Mountain. Understanding the geologic time table and its relationship to plate tectonics can provide the reader with a sketchy concept of how the landforms can be so diverse within a small region.

3. ROCK TYPES: Understanding the three basic rock classifications is an important element in being able to comprehend the formation of Dillsburg's economic mineral deposits. Igneous rocks are those "fire-

born" molten masses developed deep beneath the surface. They are generally coarse-grained, dense, dark colored rocks which can extrude to the surface by forcing through existing rocks or by being ejected violently through volcanos. Sedimentary rocks are those vast beds of rocks deposited as the heavy material of sand, mud or remnants of life at the bottom of ancient seas. Sedimentary rocks tend to be lighter in color and are often bedded or banded in appearance. Limestone and sandstone are the more common forms of sedimentary rocks. The third category is that of metamorphic rocks. This classification represents an altered form of one the first two classes, brought about by physical geological processes. For example great pressure brought about by the continual deposits of newer rock, can cause shale to alter to slate or limestone to alter to marble. A clue to the complex geological formations in the Dillsburg area is indicated by the fact that all three rock classifications can be found in the are

The following outline provides the reader with the varied rock types found in the Dillsburg area. These are noted below with the geologic landform in which they occur.

(A) South Mountain Landform

Sedimentary
Sandstone
Shale
Limestone
Dolomite

Metamorphic
Quartzite

(B) Cumberland Valley Landform

Sedimentary
Limestone
Dolomite

Metamorphic
Schist
Slate

(C) Triassic Upland Landform

Sedimentary
Sandstone
Shale
Limestone
Marble

Metamorphic
Conglomerate
Hornfels

Igneous
Diabase

CHAPTER TWO

ECONOMIC MINERAL DEPOSITS

Although there are a number of interesting minerals which occur in the Dillsburg area, this report is focused on the mineral deposits having an economic value and the history of the attempts to recover them. These deposits fall into four categories: Iron, Clay, Limestone and Copper. Following is a brief synopsis of the mineral occurrences that were sought for their commercial value:

IRON

Although iron is one of the most common metallic minerals in the earth's crust, forming a large number of minerals, only four are suitable for mining worldwide. These four ore types are of sufficient purity in iron content and volume to render them desirable for mining and processing. The four types are magnetite, hematite, limonite and siderite. The first three are forms of iron oxide and the fourth is an iron carbonate. All four types have been found in abundance in Pennsylvania and subsequently mined. The first three ore types, magnetite, hematite and limonite have, been mined in the Dillsburg area. Specimens of all three types of ore have been recently collected by the mines committee and added to the NYCHAPS collection.

Magnetite is the highest grade and most valuable iron ore found in the area. In appearance, it is a heavy black semi metallic ore which produced an iron content of 55 to 60 percent. The ore deposits occur east of the town in the area know as Ore Bank Woods. In this area the magnetite was formed as the result of a chemical reaction which took place when intruding igneous diabase in molten form came into contact with older area rocks. The lava intrusion emanating from a deep seated molten reservoir was forced upward into the surrounding country rock. This rock in the flatlands east of town consisted of sandstone and sandy limestone. These same igneous intrusions created the prominent land features of Round Top, Flat Mountain, Nell's Hill and Blair Mountain. The diabase, a silica based igneous rock was rich in iron. The contact with the sandstone and sandy limestone caused a chemical and physical reaction which released iron rich solutions from the diabase. These rich solutions concentrated into pods and veins in the country rock. Much of the country rock was altered by heat to form a harder porcelain-like rock known as hornfels. This phenomena can clearly be observed in the road-cut a mile north of Rossville on Old York Road. The diabase is the heavier grey-black rock found when traveling north from Rossville. The paler grey to tan hornfels, adjacent to the diabase, is seen near the top of the hill.

The ore pods and veins of magnetite occurred generally below the surface of the soil and were not easily seen. Discovery of this valuable ore deposit was made by Abraham Mumper in 1847, while cultivating his fields. Eventually more than a dozen mines were developed within a one square mile area south and east of Dillsburg. More than one and one half million tons of high grade magnetite ore was recovered during the sixty year period following its discovery. Large deposits of magnetite ore were also mined in eastern Pennsylvania at Pottstown, Montgomery County; French Creek, Chester County and at Cornwall, Lebanon County. The Dillsburg occurrence was the largest magnetite deposit west of the Susquehanna River. Smaller, but geologically similar occurrences were mined near Grantham, Cumberland County; Wellsville, York County and Cashtown, Adams County. The Dillsburg magnetite ore contained a large volume of pyrite (iron sulfide) and had to be roasted to remove the sulfur before it could be processed.

Hematite, another important iron ore, is found in vast beds in Minnesota's Mesabi Range and at Birmingham, Alabama. In our area hematite was found as a minor silvery, bright metallic ore with magnetite at the Mine Bank area of Wellsville.

Limonite is the most abundant iron ore in the Dillsburg area. This heavy, brown semi metallic iron ore was found along the margins of the Cambrian dolomites and quartzites of the South Mountain. Many limonite ore banks (open cut mines) were in operation in the general vicinity for decades before Dillsburg was founded. Several prominent banks were located in the Boiling Springs and Pine Grove regions of Cumberland County. Evidence of the large limonite beds located in the Dogwood Hollow west of Dillsburg on the slopes of the South Mountain were known as early as 1840. The deposition of limonite was a much more placid and time consuming process than that of magnetite.

Siderite, the carbonate of iron, has not been found in the Dillsburg area. Siderite is a low grade iron ore which has been found in Pennsylvania, primarily in association with coal bearing rocks.

Geologists conjecture that one formation of the South Mountain region known as Tomstown Dolomite, was very rich in pyrite, the sulfide of iron. Over the vast millennia, a process of chemical destruction of the pyrite (much like rusting) took place. Descending ground water solutions carried the iron downward where it was redeposited on the impervious quartzite beds which lay below the dolomite. The size of the resulting iron deposits were determined by the amount of pyrite in the dolomite and the catch basin which received the descending iron bearing solutions. Similarly, the free sulphur ions released from destruction of the pyrite, reformed as mild sulphuric acid. The acid solutions altered a certain portion of the dolomite which is a carbonate and prone to chemical destruction. The residue was reformed as a high grade of clay which had an economic value. This explained why iron ore and clay beds occur in close proximity to each other in the Dogwood Hollow area.

The two main iron ores of the Dillsburg area were formed in distinctly different environments. The random and concentrated ores east of town formed in veins and pods. Recovery of this ore began as open cut pits, but soon changed to deep shafts and tunnels that followed the ore downward. In the Dogwood Hollow, the large water borne sedimentary beds of iron oxide allowed large open cut pits for recovery. This was an important item because the limonite was of a lower grade ore which processed only 35 to 40 percent iron. Open cut mining was less costly than underground mining, and helped offset the lower percentage of iron in the limonite. Limonite, because of its abundance, was the mainstay ore of Pennsylvania's vast iron industry during the eighteenth and nineteenth centuries. As transportation improved in the latter part of the 1800s, the vast reserves of the Mesabi Range began to overtake, economically, the smaller limonite reserves. By the beginning of the twentieth century, most of the local ore banks could no longer compete. Improvements in technology and transportation brought a quiet end to a once bustling industry in the countryside surrounding Dillsburg.

This industry managed to exist a few years longer, but under-capitalization and mismanagement also brought it to a close.

CLAY

The use of clay can be traced to the dawn of civilization. The most valuable property of this mineral substance lay in the ability of early man to shape it into useful items such as vessels and bowls. Its value was further enhanced by the knowledge that it could be vitrified (hardened) by the proper application of intense heat. The art of clay mining and processing was already well established when the first deposits were discovered in Dogwood Hollow. Although no date for the discovery of the clay deposits has been recorded, its proximity to the vast residual iron beds indicate a period between 1865 and 1870.

The post Civil War era brought about a renewal of business and commerce for a war weary nation. Mineral exploration was active in the region, and along with extensive iron development in the South Mountain, an active clay industry had begun to develop in the vicinity of Mount Holly Springs, Pennsylvania. Thus, the discovery of clay in geologically similar horizons near Dillsburg was not a surprise.

The Dogwood Hollow clay occurred in three distinct varieties. White and yellow clay were the most abundant, but small deposits of black clay were also encountered. Appalachain geologist George W.

Stose made a throrough study of the South Mountain Clays in the summer of 1906. He reported that the local product was pure and remarkably free of contaminants. The white clay had greater value because its color and fine grained texture made it suitable for the manufacture of quality paper, white paint and chinaware. Yellow clay was deemed suitable for the manufacture of brick and was eventually used for this purpose. Although the yellow brick product never attained the popularity of red brick, some fine dwellings made of this material may still be seen in the Borough of Dillsburg. The less common black clay, which received its color from the inclusion of manganese oxide, was used for pottery.

Clays are generally formed through the physical or chemical alterations of pre-existing rocks. The clay in Dogwood Hollow was believed to have been derived from the disintegrated sericitic fine grained sandstone of the Antietem Formation. This hard, white material forms the cap rock of the South Mountain in the vicinity of Dillsburg. Geologists concede that some of the clay may have been transported and redeposited with the residual iron. That would provide one explanation of why the two mineral substances occurred in association with each other. Investigation of the clay deposits of the district by Stose indicated a clear relationship between the white clay and the mountain sandstone. As this relationship was recognized by the people seeking to exploit the mineral wealth of Dogwood Hollow, large deposits of clay were discovered. For the economic health of Dillsburg, the timing was good. The local clay industry began its growth about the same time as the iron industry was declining. Clay mining activities extended the life of the mining industry in the Dogwood Hollow district for several more decades.

The exact process which formed the clay cannot be completely understood. As one views the wooded summit of the South Mountain, standing more than 500 feet above the surrounding area, it is difficult to conceptualize this feature as the bed of an ancient sea. The bottom of any body of water will accumulate the sediment which drifts downward. If given sufficient time, sea bottom sludge which consists of mud, sand, the exo-skeletons of sea life and other refuse, will compact and become rock. The South Mountain rock plate, once the bed of an ancient ocean, was violently broken and uplifted by continental drift in geologic prehistory. Though the rock plate consisted primarily of compacted sand, some remnants of limestone and micaceous slate were also included. Later, chemical and physical alteration caused some margins of the rockplate to disintegrate and reform as a plastic-like clay mass; the residue of a mix of sand, lime and tiny flakes of included mica. The randomly formed mineral substance remained intact through the geological ages until its discovery in the latter part of the 19th century.

COPPER

The need for copper grew dramatically during the latter part of the nineteenth century. Though no major copper occurrences existed in the northeastern United States, copper mineralization is abundant and very colorful. The bright and profuse mineralization has spawned more abortive attempts to mine and process copper. Efforts to develop the copper industry were greater than those of any other base metal. Dillsburg had its share of these attempts. Copper is a minor constituent of the local rocks of the Triassic Uplands. The intruding diabase caused the copper, like the iron ore, to be released and redeposited as a minor mineral deposit. There proved to be no major copper occurrences in the Dillsburg area. A local example of copper mineralization can be seen at the previously described roadcut near Rossville. The colorful copper carbonates, malachite and azurite, can be seen on the hornfels along the east side of the roadcut.

LIMESTONE

Limestone, a common sedimentary rock, is familiar to most people. Outcrops of white, tan, gray and bluish-black limestone are plentiful throughout Pennsylvania. Because limestone is easily eroded, it comprises the base rock of many low lying areas such as the

Cumberland Valley, Lebanon Valley and York Valley. Its economic use is also well known. To the early settler in the American colonies, it had a number of important uses such as soil conditioning, flux in iron smelting and even whitewash.

Limestone occurrences near Dillsburg were neither high quality nor abundant. As a result, no significant quarrying or processing of limestone was developed in the region. Most of the deposits, which lay under the town and westward to the base of the South Mountain, were in the form of limestone conglomerate. This rock consisted of massive amounts of waterworn pebbles and boulders bonded together by limestone. Conglomerate is a peculiar rockform which developed during the traumatic mountain building process. This process created Dillsburg's most prominent landform. Massive amounts of rock that was broken up and fragmented during the tectonic coastal breakup was washed to the base of the mountain as allurium. The alluried deposit was cemented together by sediment that followed as the lowlands were covered by an invading sea. Subsequent pressure compacted the material into the rock which lay undisturbed until the first settlers arrived in the region.

It is likely that burning (dehydration) of limestone began as early as the 18th century. State geologist Persifor Frazer (1877) reported that Dillsburg native John Kuntz had a profitable business in the quarrying and burning of the Triassic limestone conglomerate during that decade. Kuntz operated his quarry and kilns about one mile west of Dillsburg near the slope of the South Mountain. Many years later, Strayer and Lefever of Dillsburg operated a quarry near Williams Grove. The rock at this location was identified as Tomstown dolomite, a high magnesia limestone which was suitable for road building material. The Tomstown formation was considerably older than the conglomerate, and was identified as belonging to the Ordivician series of limestones that comprise the Cumberland Valley. Although limestone production in the Dillsburg area was discontinued many years ago, the 700 feet long, 60 feet high Williams Grove quarry is still a prominent landmark and holds mute testimony to Dillsburg's mining history.

SECTION I

BIBLIOGRAPHY

Bloom, Robert L., Prof. Emeritus of History, Gettysburg College - *A History of Adams County, Pa. 1700-1990.* (1992)

Bricker, Gilbert - Interviewed by Bob Gayman and John Livingston. Spring 1992.

Dillsburg Vitrified Brick and Tile Company official documents - Corporation book and correspondence. 1906-1911.

Fishel, Leo - Telephone interview by Bob Gayman, January 6, 1993.

Flower, Lenore Embick - "Central Iron and Steel Company, Harrisburg, Pa. - History of the Plant."

Fortney, John C. - *The Gazette and Daily, York, PA,* October 27, 1933, p.15 - John Morris, Edison's Mining Expert, Former Dillsburg Man.

Frazer, Percifor, Jr. - "Report on Progress in the District of York and Adams Counties", *Second Geological Survey of Pennsylvania,* (1874) pgs. 71-76.

Frazer, Percifor, Jr.- "Report on Progress". *C235-237 The Dillsburg Group of Mines,* (1875).

Frazer, Percifor, Jr. -"The Dillsburg Group of Mines. Ore Belts of York, Adams and Cumberland Counties Pennsylvania".

BIBLIOGRAPHY

Frederick, Lewis - Interviewed by Lee Karns.

Geological Survey of Pennsylvania Annual Report for 1886. Part IV, pgs. 512-1514.

Geyer, A. - "Mineral Collecting in Pennsylvania, G-33, Dillsburg Iron Mines."

Gibson, John - *History of York County*, (1886), pgs. 649-651,

Hain, H. H. Marshall Furnace, *History of Perry County, Pennsylvania,* (1922), pgs. 279, 1922

Harder, E. C. - "Structure and Origin of Magnetite Deposits near Dillsburg, York County, Pennsylvania" *Economic Geology*, (Oct-Nove 1910), pgs. 599-622, Oct-Nov 1910.

Hinchliff, Helen, family historian of Mumper family - correspondence Dec. 4, 1990.

Himes, Ed. - Note of personal recollections on black clay mine. No date given.

History of that part of the Susquehanna and Juniata Valleys, embraced in the Counties of Mifflin, Juniata, Perry, Union and Snyder in the Commonwealth of Pennsylvania, Volume II - Marshall Furnace, Juniata Furnace, p. 1090.

Hoff, Wilma Weldie - Telephone interview Nov. 25, 1991 re: Z.W. Manufacturing.

BIBLIOGRAPHY

Jones, Jeri L. and Goodman, Jan - "The Iron Mines of Dillsburg, Pennsylvania". *Rocks and Minerals* July - August 1980.

Karns, Lee - Notes/interviews 1992.

Kelley, Mrs. Arthur (Helen) - Interviewed by Jack Winieski, 1994

King, Clarence - Interviewed by Lee Karns, July 26, 1985.

Kreiger, Bill - Interviewed Jeri L. Jones. Work to Catalogue County's Mines, York Daily Record, Dec. 7, 1986.

Leighton, Henry - "White Clays of Pennsylvania. Dillsburg Region, York County", *Pennsylvania Dept. of Internal Affairs, Bulletin 112,* (July 1934), pgs. 14.

Miller, Benjamin Leroy - "Limestones of Pennsylvania". *Pennsylvania Geological Survey Fourth Series, Dept. of Internal Affairs, Harrisburg, Pa.*, (1934) pgs. 713-714.

Miller, Diane R. - "Early Brick Factory Beset By Money, Supply Problems". *The Weekly Bulletin,* (June 3, 1981).

Mumper, John - Ledger 1841 - 1864 - donated to Pennsylvania Archives.

Pennsylvania's Mineral Heritage, York County, *Pennsylvania Dept. of Internal Affairs,* (1944).

Prowell, George - *History of York County*, Vol. I, pgs. 939 also pgs. 1027-1028.

BIBLIOGRAPHY

Richley, John William - Autobiography titled *Obstacles No Barrier*, York, Pa. Lithographed Printing Plate Craftsmen, (1951).

Schlusser, Mildred - Interviewed by Jack Winieski, 1994.

Schrack, Susan, local resident recounts mining disaster (Luke Haar). *Weekly Bulletin*, (?).

"Thomas A. Edison, The Great Inventor.....", *Dillsburg Weekly Bulletin,* (August 4, 1905).

Shambaugh, H. J. - Interview summer 1992.

Slough, David E. - Interviewed by Lee Karns Jan. 20, 1986.

Spencer, Arthur C. - "Magnetite Deposits of the Cornwall Type in Pennsylvania". *Bulletin 359, U. S. Geological Survey*, (1908).

Spencer, Arthur C. - "The Jauss Iron Ore Mine, Dillsburg, Pennsylvania". *Contributions to Economic Geology,* Part 1, (1909).

Steinmetz, R. H. - "Industry Once Thrived Here". *The Weekly Bulletin,* (Mar. 3, 1982).

Steinmetz, R. H. - "Railroad Once Linked Dillsburg, Mechanicsburg". *The Weekly Bulletin,* (Mar. 3, 1982).

BIBLIOGRAPHY

Steinmetz, R. H. - "19th Century Plans Might Have Made Dillsburg a Railroad Center". *The Weekly Bulletin*, (Sept. 23, 1986).

Stose, George W. and Jonas, Anna I. - "Geology and Mineral Resources of York County". *Bulletin C-67. Pennsylvania Geological Survey, Fourth Series,* (1939).

Stough, Carla; Kirkland, Sue and Prober, Paula - "Some Memories of Old Dillsburg: Past is Gone but Not Forgotten". *The Patriot*, (Aug. 18, 1980).

U.S. Department of Interior, Bureau of Mines, *Investigations of the Dillsburg Magnetite Deposits, York County, Pa.*, (Dec. 1947).

Watts, Randy - *Railroads of the Dillsburg Region.* Book three in a series, Railroads of the Cumberland Valley, (1992).

Weaver, Cleason John - Interviewed by John O. Hoffman, Sept. 16, 1973.

Weaver, Rev. William H., first pastor of Calvary United Brethren Church, East Harrisburg Street, Dillsburg, Pa. 1894-1899 - paper written around 1884.

Westhaeffer, Paul J. - *History of the Cumberland Valley Railroad, 1835-1919.* " Chapter X - The Iron Ore Branches." Washington D.C. Chapter National Railway Historical Society, (1979).

SECTION II

SELECTED READINGS

Frazer, Persifor, Jr. -"Report of Progress in the Counties of York, Adams, Cumberland and Franklin". *Pennsylvania Geological Survey,* (1877), pgs. 201-400

Geyer, Alan R.; Smith, Robert C. II and Barnes, John H. - "Mineral Collecting in Pennsylvania", *Pennsylvania Geological Survey G33,* (1976), pgs. 218-221

Harder, E. C. - *Structure and Origin of the Magnetite Deposits near Dillsburg, York County, PA . Economic Geology Vol. 5, No. 7,* (1910), pgs. 599-622

Holtz, Preston E. - "Diamond-Drill Exploratin of the Dillsburg Magnetite Deposits, York County, Pennsylvania". *U.S. Geological Survey Bulletin 969A,* (1950), 25 pages

Jones, Jeri L. and Goodman, Jan - "The Iron Mines of Dillsburg, Pennsylvania" *Rocks and Minerals Magazine Vol. 55, No. 4* (1980), pgs. 153-157

Miller, Benjamin L. - "Limestones of Pennsylvania", *Pennsylvania Geological Survey Bulletin M20,* (1934), pgs. 336-346

Neumann, G. L. - "Investigation of the Dillsburg Magnetite Deposits, York County, PA" (1947). *U.S. Bureau of Mines R.I. 4145, 13* pages

SELECTED READINGS

Smith, Robert C. II; Berkheiser, Samuel W. Jr. and Hoff, Donald T. "Locations and Analyses of Selected Early Mesozoic Copper Occurrences in Pennsylvania". *U.S. Geological Survey Bulletin 1776*, (1986), pgs. 320-332

Spencer, Arthur C. - "Magnetite Deposits of the Cornwall Type in Pennsylvania" . *U.S. Geological Survey Bulletin No. 359,* (1908), pgs. 71-96

Spencer, Arthur C. - "The Jauss Iron Mine, Dillsburg, Pennsylvania" *U.S. Geological Survey Bulletin No. 430,* (1910), pgs. 247-249

Stose, George W. and Jonas, Anna I. - "Geology and Mineral Resources of York County, PA". *Pennsylvania Geological Survey Report C67*, (1939) 199 pages

Wherry, Edgar T. - "The Newark Copper Deposits of Southeastern Pennsylvania". *Economic Geology Vol. 3, No. 8*, (1908), pgs. 726-738

INDEX

INDEX

INDEX

INDEX

INDEX

INDEX

INDEX

INDEX

INDEX

INDEX